BUILDING
THOREAU'S CABIN

by

STEPHEN TAYLOR

PHOTOGRAPHS BY
KEN ROBBINS

PUSHCART PRESS, WAINSCOTT, NEW YORK

Copyright 1988, 1992 by Stephen Taylor and Ken Robbins
All rights reserved.
For information address Pushcart Press
Post Office Box 380, Wainscott, New York 11975

ISBN 0-916366-74-X

Previously published as:
A PLACE OF YOUR OWN MAKING
(Henry Holt and Company, 1988)

Distributed by W.W. Norton & Co., New York

For Barbara Hulsart and Maria Polushkin Robbins

This might never have happened without the advice and support of Charles Eric Engstrom.

Our culture is the predominance of an idea which
draws after it this train of cities and institutions.
Let us rise into another idea; they will disappear.
                —*Ralph Waldo Emerson*

Contents

    Introduction   1
1. Preparations   13
2. Making a Sketch   27
3. The Foundation   36
4. The Deck   46
5. About Windows and Doors   75
6. Framing the Walls   81
7. Framing the Roof   102
8. Sheathing, Siding, Windows, and Door   115
9. Closing the Roof   129
10. Ingress: Decks and Stairs   144
11. When to Stop   155
12. Power   160
13. Insulation and Vapor Barrier   177
14. The Interior Walls   184
15. Completing the Room: Trim and Flooring   198
16. Paint, Heat, Finishing Touches   217

    Glossary   233
    Index   239

Introduction

"The purpose of imagination," wrote Martin Buber, "is to imagine the real."

Exploring the implications of that tantalizing, multilayered aphorism could easily be the subject of an entire book, if not a life's work. Yet building a small building—not reading about doing it but actually building the building—can be, in its special way, another means toward accomplishing some of the same thing. I'm not saying you ought to build an outbuilding in order to understand more thoroughly what Martin Buber meant (although that might not be a half bad idea). But if you do decide to go ahead and build, you may find that, indirectly, Buber will be offering at least as much help as you'll get from me.

I should explain what I mean. Most of us spend a sizable portion of our lives surrounded by wires and windows and heating ducts and pipes, by structural support members fashioned of wood or metal or concrete fastened with nails or rivets or bolts, by sheets of plaster and plastic and plywood and heavy paper soaked in tar, perhaps by tiers of bricks or mosaics of shingles or matrices of boards, and, strangely enough, we call this condition "being at home." Of course, when we *are* home, we seldom take the trouble to imagine how all those materials fit together. As long as they leave us alone, we're usually more than happy to leave *them* alone. But sometimes one or more of those materials assert themselves. They rot or burst or spring a leak or otherwise malfunction, and suddenly we're called upon to give them our attention. All along they've been quite real, but their reality's been passive, concealed behind the scrim of God only knows what other competing concerns. Now, abruptly, their reality springs to life and makes us oddly anxious.

Probably the anxiety we feel is of two kinds. The first kind has to do with all the questions of how bad the damage is— how much it will cost to fix, how soon it can be got to, how uncomfortable or inconvenienced we'll be until it's taken care of; all the obvious considerations that ensue when something crucial doesn't work. But the second component of the anxiety is much more vague and maybe just as troubling. It's the component of mystery, of things no longer being in control. What's gone afoul inside the walls? Just what is it that's happening in there?

You want to allay your anxiety, so, perhaps habitually, you say to yourself, "I'll call someone." Whereupon you find a carpenter or plumber or electrician or handyman—someone to whom imagining the reality of walls like yours is all in a day's work—and, if you're very lucky, he comes reasonably soon, fixes what ails your house reasonably well, and charges you a reasonable fee. He thereby puts the first kind of anxiety, the kind that comes of things not working right, to rest. But the second kind, the anxiety engendered by the mystery inside the walls, by the fact that there are things nestled in the darkest recesses of your house that you may not be able to fix or understand or even picture, persists. Now you don't feel quite as much at home when you're at home. Yes, your house is real enough and really yours, but you've had to hire someone else to imagine its reality. When you called in a repairman, what you really did was to take an objective problem, a problem with things (in this case with your house), and transform it into an interpersonal problem. Instead of dealing with what went wrong with your house, you wound up dealing with a person who could fix it. You jollied him into coming promptly, offered him beer or coffee when he got there, negotiated a price that both of you could live with, and maybe sized him up in terms of establishing an ongoing working relationship. But as for *your* imagining the reality of your house, you ducked the issue altogether.

Many of those writers and thinkers of the first half of this century who came to be called existentialists, Martin Buber among them, dealt at length with the question of our feeling less at home in the world than perhaps we once did. The world, they said, has become increasingly unfathomable, its reality harder and harder to imagine. In part this was ascribed to the burgeoning complexity of the world, but some of the blame had to be placed with the people who inhabited it. Often we no longer *try* to imagine the world, to imagine, as Buber put it, the real. Whenever the real needs imagining—and it certainly needs to be imagined by a plumber trying to guess where to chop into your wall to find a leaky pipe—we too often pass the job along to someone else.

Now, it's not my intention here to put repairmen out of business. The world's assuredly complex, and if we were to refrain altogether from calling upon specialists its complexity would quickly overwhelm us. If I fall off a ladder, crack a rib, and then manage to drag myself over to my doctor's office, I'm going to be less than happy if his secretary tells me that he's home tuning up his car. But specialization can also be abused, and habitually resorting to specialists can become a bad habit. Consider the 1950s stereotype of the father as a man who might be a crackerjack at the office but couldn't tell child care from a cheese sandwich and was equally clumsy with both. It took a small social revolution to encourage him to imagine the reality of children and sandwiches and eventually the reality of people who *deal* with children and sandwiches, and his initiation into those realities was probably accompanied by considerable kicking and screaming. But one day, puttering in the kitchen, he realizes that instead of feeling helpless and out of place, he's actually very much at home. Where he was once an oaf he's now an adept. He hasn't renounced all specialization, but he's no longer quite so addicted to it. His known world has become measurably larger.

What I'm suggesting is that far from encumbering your mind

New walls

with knowledge you'll never need, exploring the mysteries inside your walls—and doing it by creating a whole new set of walls from scratch—may free you to ponder much greater and more important mysteries later on. I'm suggesting, idiosyncratic as this might sound, that building your own outbuilding is a way back to your home, to the real, to the world.

Whether elevated to the rank of "studio" or demoted to the homely designation "shed," the outbuilding has quietly begun to make a comeback, and the fundamental premise of this book is that nearly anyone can build one more or less by himself. I don't mean a flimsy, amateurish junk pile either, although back when they were a commonplace on rural properties, it's precisely as flimsy, amateurish junk piles that thousands of American outbuildings were born. Until fairly recently, country folk seldom stopped to wonder if they were capable of putting up a small building. They simply went ahead and did it. If the resulting structure was more than a little ramshackle, its improvement was left to future generations. Unless, that is, it fell down before future generations even got to it. Rural New Englanders used to explain why their town hadn't adopted any formal building code by saying, "What's the point? If a buildin' ain't no good, the snows'll take it."

Nonetheless, the outbuilding we'll be describing here is a sturdy, weathertight structure that's eminently usable, decently presentable, and capable of passing muster with just about any building inspector. Given some guidance and a modicum of hand-holding (both of which, first time around, are likely to be at least as important as "how-to" instruction), an unskilled layman can expect to put up such a building in a surprisingly short time and for surprisingly little money. And when he's done, he'll have the bargain hunter's private delight in getting what he got for considerably less money than he'd have paid a specialist (read "builder") to do the job. The value of his property will have increased far out of proportion to what the outbuilding actually cost, and he'll also have had the chance to be outdoors regularly and for a good reason. He'll even have had some occupational therapy of the best sort, in that he really wanted the end result. He'll have the pride of accomplishment that ensues from building something substantial, useful, attractive, and enduring. For a time he'll have had—and the importance of this isn't to be minimized—a fresh obsession to replace the ensemble of old obsessions that normally beleaguer him. And he'll wind up with the pungently subversive pleasure that comes of successfully practicing a craft without taking a whole lot of time or effort to master it.

Which brings up the question: if you're going to build a building and you don't have much craftsmanship to draw on, what *do* you have to draw on? Well, it's here that I'd like to touch on the concept of field expedience. If this were a conventional how-to book, it would bear more than a passing likeness to a recipe book. Before beginning to tell you "how-to," it would rigorously list the tools and materials you'd need to do the job. Yet the stance that proper tools are essential to a task is one the novice can choose to adopt or leave alone. A craftsman who keeps making the same things again and again, who's deeply wedded to long-held routines, would understandably be lost without what he feels are the proper tools. The beginner needn't be that hidebound. Not being a craftsman, he's not obliged to behave like one.

But, lacking craftsmanship, the novice builder is by no means without resources. He can still put his trust in ordinary competence, ordinary good sense, and ordinary ingenuity. If craftsmanship develops over time, competence, good sense, and ingenuity can be tapped right now. When a board has to be sawn to a given length, you can do it with an everyday hand crosscut saw, a hand backsaw, an electric circular saw, a table saw, a band saw, a radial-arm saw, a power miter saw (sometimes called a chop saw), a hacksaw, a saber saw, or the little sawblade in a Swiss Army knife. (If you're unfamiliar with tools—especially power tools—take a look at a Sears catalog just to see what's around.) Used correctly, any one of those will get the job done. It's true that some will get it done faster than others, or more easily than others, or with more probable precision than others. But they're all capable of doing it, and even apart from cost, it may take considerably less time to cut a board with the "wrong" saw that's close at hand than with the

Cutting boards with a circular saw

proper saw that you have to make a round-trip to the hardware store to buy. So, in the name of field expedience, you really don't have to assume that any tools other than a decent claw hammer, an ordinary level, a line level, a tape measure, a standard $7\frac{1}{4}''$ circular saw with an all-purpose blade in it, and maybe a nail-puller and an inexpensive caulking gun are essential to this project. I'll mention other tools as we go along, but the chances are excellent that whatever household tools you have around will suffice to pick up the slack. A hefty stone can accomplish many of the same things that a framing hammer can, and one day when you can't remember where you've left your hammer you may, in a fit of field expedience, find yourself reaching for a stone. Craftsmanship can sometimes be an impediment to ingenuity. Field expedience encourages it.

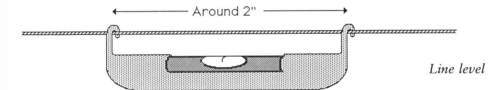

Line level

Where the experienced carpenter does have a clear edge over the novice, however, is in the crucial area of terminology. A seasoned builder can walk into a lumberyard or a hardware store and reel off a shopping list that, to a beginner, might easily sound like so many ravings in medieval Ruritanian patois. And it's here that a book can be very useful to the tyro. Watching the guys on line ahead of you be treated with deference and respect simply because they appear to know what they're talking about only to be patronized nearly to the point of pity when your own turn at the counter comes is an experience there's every reason to avoid. When you build the outbuilding, there'll be enough dues to pay without having to be hazed in the bargain.

Still, it's not on public occasions but during those building sessions when the only discourse is between you and the lumber that the real power of terminology becomes most apparent. Having names for things is one way a carpenter puts himself at ease with them. It's not hard to imagine such a carpenter staring warily at a narrow strip of sheet metal in a lumberyard and slowly working up to asking the clerk, "What's that?" If the clerk expects to keep his job, one reply he won't make is "Are you kidding me? You don't know what that is?" Knowing what things are is at the heart of any seasoned builder's self-esteem, and the clerk would be far more likely to say something like "Aahh, it's just some new kinda drip edge supposed to stand up better to cedar fascia that's got a lotta sap in it." The builder would know exactly what the clerk means, and so, if you build the outbuilding, will you. Because the builder's own internal glossary of terminology has let him easily capture some new bit of knowledge, he feels more comfortable now, and so, as your own construction glossary grows larger, will you.

For the first several sessions you may not be very fluent at assigning names to all the tools and materials you're using, much less to the processes you're carrying out or the intermediate results you achieve. But then, as you begin using your tools and materials and this book more and more, as you begin using *yourself* more and more, your repertoire of terminology will grow. And the more fluently you can name things, the more secure you'll be in your possession of them. After all, if you own a house, what you have is just a house. But if the house has soffits, and you know what soffits are (maybe because you've worked on some), you also own soffits which, in a sense, you wouldn't otherwise own.

The message is simple. By acquiring some new terminology in the course of building the outbuilding, you wind up "having" more things without physically increasing your inventory of possessions. The "having" is interior. It's in you. And that's another reason to go ahead and build. Yes, when you're all done you'll have a small building that will serve whatever private purpose you have in mind. But a building's just a building. The real increase in your assets will be in your imagination. A word as simple as "roof" may come to mean something much different than it did before. It may conjure images of rafters, plywood decking, ridge and soffit vents, overlapping layers of roofing felt, nailheads carefully left unexposed—of disparate materials joined by a variety of techniques to compose a functioning entity. Suddenly this mundane monosyllable—"roof"—will resonate in your imagination as it never did before. And we've already noted that the purpose of imagination is exactly that: to imagine the real.

Of course, now that you see the direction we're taking, you may not *want* to have as ordinary a word as "roof" explode so disconcertingly every time you hear it. You may prefer not to know what's inside a roof, or, for that matter what's inside anything else. You may feel that having "roof" reverberate that way will bombard you with more mental clutter than you care to contend with. The inner truth of roofs—of anything—may be repugnant to you. You may be perfectly pleased with an imagination that hovers near the surfaces of things and leaves their innards for others. You may, in a word, be shallow and even want to stay that way. There are circles—shallow ones, to be sure—that endow certain kinds of high-toned shallowness with an aura of chic, and you may aspire to be a member of such a circle, if you aren't one already. Unfortunately, there's nothing terribly chic about building an outbuilding, and if you routinely see construction details as mere clutter and would just as soon not change, this may be a good time to stop reading and pass the book along to someone else.

If, on the other hand, you're intrigued by the sort of change I'm speaking of, building the outbuilding may be a step toward bringing it about. You set forth with a simple enough wish: to have an outbuilding or studio or shed or cabin or detached cottage or one-room shack or whatever you want to call it and maybe to save some money and get some exercise in the bargain. But as the structure starts going up, the nature of your quest evolves. The building proceeds to change the builder. As your manual skills develop, your confidence swells. Materials which, only a few days ago, were a little awesome and intractable are now familiar and obedient. There's a growing harmony between you and your tools. On a given day they seem able to do more things than they could the day before. Nails start going in faster and straighter and with fewer blows of the hammer. Building any building—even a little one—is no small task, but gradually you begin to feel it's a task you're more than up to. And by the time you're done—you've been to the lumberyard and the hardware store many times by now, and at both places you've long since given up your self-deprecating stammer—there comes a moment, however brief, when you're certain you could build the Taj Mahal, or at least its rustic equivalent in wood. You've hit a new plateau, and you can't keep from patting yourself on the back.

I mentioned the dues you might have to pay to come as far as that, but they're hardly exorbitant. People who talk about changing in some sought-for way will insist on their "no pain, no gain" caveats, yet there's something in the very nature of the process of putting up a building that makes the evolution of your skills comparatively painless. Step by step, your proficiency grows as it's most needed. At the outset of the project, the foundation asks for very little in the way of expertise, so your ability to transfer what's in your imagination to the ground or to the wood doesn't have to be especially refined. You'll get an early feeling for your tools and materials, and the foundation will forgive all but the grossest mistakes. Then, when you go on to the deck, you'll need only slightly more precision than

The deck partly framed

you needed for the foundation, and when you finish it you may find yourself aching to use your developing facility on something more demanding. And sure enough, framing the walls, the next step in sequence, is just that thing. By the time you're in the homestretch you'll be trimming the inside of the building—making windowsills, baseboards, all the woodwork that draws the closest and most frequent scrutiny. And by then, having practiced at cruder tasks, you'll have gained enough proficiency to make the work look decently professional and maybe even ornamental. Newfound skills have a way of asking to be practiced ornamentally.

Finally I'd like to take note of the fact that whenever this book makes reference to the imagined reader, the clerk at the lumberyard, the counterperson at the hardware store, or any number of other characters apt to be encountered en route, they're all called "he." While women oughtn't in any way be intimidated by the prospect of building an outbuilding, it would be foolish to pretend that the great majority of this book's users won't be men. Still, say what you like about the sexual revolution of the past several decades, it's certainly made it practical for a woman to undertake the building of an outbuilding without being thought of, or feeling like, a freak. An oddity, maybe, but not a freak. The nail *does* know who's swinging the hammer, but what it knows is whether the swinger is hitting it squarely or obliquely, not the composition of the swinger's hormones. I'll continue to say "he" for my own convenience, but I doubt that any woman seriously inclined to build a small building will let herself be flustered by a pronoun. My assumption throughout much of the book is that the reader will be building alone. However, where it's unsafe or impractical to work alone, I'll suggest that help be found. And if good help becomes available at other times, even when safety isn't a factor, it would be unwise not to grab it. Materials can be very heavy, for men as well as women, and help that's truly help is always welcome. I hope this book—and Martin Buber too—will be of help.

1

Preparations

Most professions have their long-held truths, and one in the building trades is that no house is ever really finished. As a house goes up, its owner typically keeps getting new ideas, and the builder, after calling most of them impractical or too expensive, selectively incorporates a few. Finally, when the house is more or less habitable and whatever shared enthusiasm there's been between client and builder has degenerated into animosity, the builder tires of the project and stops showing up, the owner relaxes his standards and decides to make do, and the absence of momentum comes to be interpreted, willy-nilly, as completion.

The process of designing a building is seldom really finished either. It's apt to begin as a surge of wishes and fantasies that precede even the first crude drawing, and, long after the building is up, it tends to tail off as a series of melancholy afterthoughts riddled with phrases like "If only I'd . . ." and "Maybe I could still . . ." and "What would it take to . . . ?" Somewhere in the middle of all that comes the creation of a working sketch, but it's virtually essential to do some constructive daydreaming first. Wishes and fantasies are fine, but while you're having them you might also make yourself comfortable with the principles of house construction, give some thought to site selection, and explore with your building department whether putting up an outbuilding on your property is even legal.

You may also want to do some realistic deliberation about allocating time and money—let's say, very *very* arbitrarily, a dozen or more weekends and a couple of thousand dollars. And, yes, there are all sorts of ways you can shave money off the construction budget; as with most things, time and money have their usual inverse relationship. If, for instance, there's a dump nearby, you may be able to scavenge some perfectly usable building materials. I met someone not long ago who made it a point of honor to build an outbuilding entirely from materials he found at the local sanitary landfill. Of course that kind of scavenging took time and patience, and I need hardly say that the materials didn't turn up in the sequence they were needed.

By and large, decisions about time, money, the size and look of the building, where on your property it will wind up, and how to please your building inspector are ones you'll have to make yourself. But decisions can be wise or foolish, and a certain amount of preparation can steer most of them toward being good ones. So if you already know how a frame house is put together and what its basic parts are called, you can either skip the next section or read it just to make sure we're talking about the same thing.

SOME ESSENTIALS OF FRAME DESIGN

Basically, a frame house is one that's supported by its walls. The walls, that is, do more than just enclose the house; they also hold it up. This contrasts with, say, post-and-beam construction, in which most of the weight of the building sits not on its walls but on horizontal beams supported by vertical posts inside the building. You commonly see post-and-beam design in barns, where the exterior walls are often too tall and flimsy to be relied on for support and where, because of the large spaces a barn has to enclose, there aren't enough interior walls to support the lofts and large roofs that barns generally have.

Sometimes residential buildings combine frame design with post-and-beam, especially if they have a very large downstairs space—the living room, let's say. The rest of the downstairs, composed of smaller rooms, would be crisscrossed with enough walls to support the upstairs (which is why those walls would be called "bearing walls"; they bear the weight of the story above). But the big living room, being an expanse uninterrupted by such walls, might have to be spanned overhead by a number of beams supported here and there by posts.

The outbuilding we'll be putting up, however, relies solely on its walls for support, so it's of pure frame design. Frame construction is probably the easiest kind for the beginner because structural members are comparatively light and easy to join together. It also requires only a minimum of engineering, and its basic principles can be mastered—make that "comprehended," since mastery is somewhat beyond the scope of this book—fairly swiftly.

We'll start from the bottom and work our way upward.

The lowermost—and normally the most massive—horizontal members of a standard frame structure are the girders. They usually run the length of the house, and in full-sized houses they're ordinarily doubled or tripled 2x12s. If the house has a concrete basement or crawl space, their ends often sit in depressions molded into the concrete called "girder pockets." And if they span any appreciable distance they're usually supported at regular intervals by cylindrical steel posts called "Lally columns." (The post-and-beam principle intruding once again.)

The boards laid horizontally along the tops of the foundation walls are called "sills" or "house sills." They're usually bolted to the foundation, and because they transfer their load directly to the foundation they're generally thinner than the girders. The top surfaces of the sills are level with the top surfaces of the girders, and for all practical purposes they function just as girders do.

Now, set at right angles across the tops of the sills and girders, come the floor joists. They'll most likely be 2x10s or 2x8s or even 2x6s, depending on how much distance they have to span. If construction is conventional, the joists will be spaced at 16" intervals. Since they're sitting on their short side (the 2" side

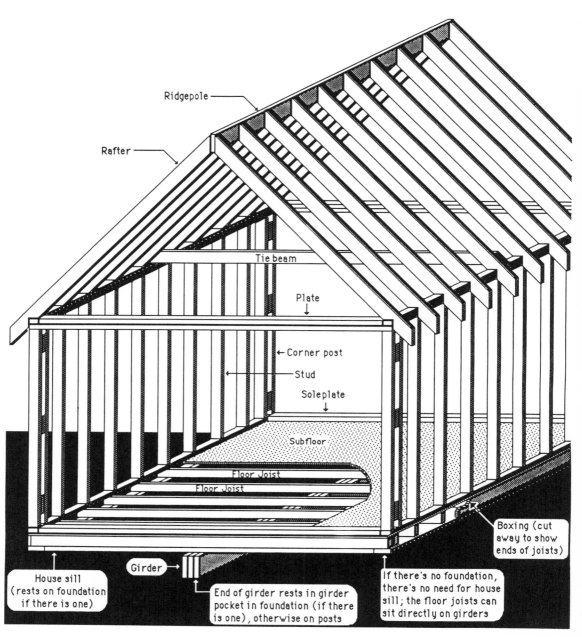

Some basic parts of a frame structure (with some pieces cut away and a couple of walls missing)

rather than, say, the 10″ side), there's the chance they could twist, so if they span any appreciable distance between girders or between girder and sill, the joists will have either wooden or metal "bridging" nailed between them. The simplest bridging consists of sections of the same size boards as the joists themselves. An alternative method is called "cross-bridging," in which wood or metal struts are nailed crosswise between the joists.

Probably the ends of the joists will be boxed in. That is, if the joists are, say, 2x8s, there'll probably be lengths of 1x8 lumber (2x8 if the builder's been generous) nailed to their ends. The top surfaces of the joists, together with the top surfaces of the boxing, form a level plane, and nailed down onto it are sheets of plywood, usually $\frac{1}{2}''$ or thicker, that constitute the house's subfloor. Taken together, the sills, Lally columns, girders, floor joists, boxing, and subfloor constitute what's called the "deck." During the building of a frame house, the completion of the deck is a milestone. Everything from here on in (except for any brick or masonry chimneys) will rest on, and therefore be supported by, the deck.

Next the walls are laid out on the subfloor. In most conventional construction—and for simplicity's sake we'll be conventional—the walls are framed with 2x4s, so 2x4s are placed on their flat side (i.e., the 4" side) around the perimeter of the deck where the outside walls will stand. More 2x4s are laid where the *interior* walls will stand. Gaps are left for the interior doors, and when all these 2x4s are in place they become an actual-size floor plan for the first story of the house. These bottom members of the walls are called "sills" too, but they're wall sills as opposed to the house sills we spoke of earlier. (They're also called "soleplates.")

Next, the vertical members of the walls, also 2x4s, are erected on the sills. These vertical members are called "studs," and they're cut to give the desired ceiling height, normally around 8'. Like floor joists, they're spaced 16" apart, and wherever possible they're placed directly above a joist. This procedure—keeping the studs *over* joists and not between them—is called "stacking," and it ensures that the upstairs floor, if there is one, will have a minimum of bounce.

Nailed horizontally across the tops of the wall studs is a double row of 2x4s called "plates" (sometimes called "top plates" or "wall plates" as distinct from "soleplates"). Sometimes, because of the doubling of the 2x4s, they're called "double plates." The plates are configured pretty much like the sills, and often what carpenters do before nailing up the studs is lay out the plates directly atop the sills to make sure they match. Then, wall by wall, they lift the plates off the sills and nail the studs up between them.

If the house has an upstairs floor, a second set of floor joists is nailed across the downstairs wall plates. The joists are supported at their ends by the exterior walls and along their lengths by whatever interior walls they cross. If they happen not to cross any interior walls and still need support along their lengths, the house may have additional floor girders to hold them up. A second subfloor is then nailed onto the upstairs joists, and now there's an upstairs deck on which to erect the second-story walls. Those walls, in turn, culminate in the upstairs plates, so, whether there's one floor or two, we still end up with a set of plates on which we have to build a roof.

The basic structural members of a frame house roof are called "rafters," and their purpose is to span the distance between the top plates of the exterior walls and what's called the "ridgepole." The ridgepole—actually a board and not a pole—is the horizontal member that goes along the peak of the roof, and the upper ends of the rafters are cut at an angle so that they butt the sides of the ridgepole cleanly. The rafters of an ordinary house are likely to be 2x6s, 2x8s, 2x10s, or even 2x12s, depending on what the distance between the plates and the ridgepole is. For several reasons, building codes permit a 2x8 rafter to span a greater distance than a 2x8 floor joist. Not being a floor, the roof of a frame house doesn't have to support pianos, washing machines, refrigerators, or bathtubs full of water. (Yes, it does have to support snow and the occasional roofer, but on a load-per-square-inch basis, both are feather-light compared to a full cast-iron bathtub.) Also, even if there's a heavy snow load, the slope of the roof—and starting now, let's get serious and refer to a roof's slope as its "pitch"—the pitch of the roof will aim some of the load downward onto the walls. Exceptions to this are flat roofs, which don't have any pitch, but, questions of structure aside, flat roofs are notoriously leaky, so for now we won't even consider them as possibilities.

Let's stay with rafters for a moment, because of all the structural members of a frame house they're the most complex. Since their upper ends are cut off at an angle other than 90°, the length of the cut is greater than the width of the board. If we're speaking of a 2x6 rafter, the end cut might be around 8″ long. Consequently, the ridgepole into which the rafters butt would be a 2x8, not another 2x6. That way the ridgepole is butted by the rafters along the entire length of their end cuts, which gives a result that's more pleasing both structurally and aesthetically.

Unlike their upper ends, however, the lower ends of the rafters seldom butt directly into the plates. If they did, the roof would have no overhang at all, and rainwater would drip down along the outside walls and windows. Instead, each rafter usually has a notch cut into its underside about a foot or so from its lower end. The notch is called a "bird's-mouth," and the two sides of the notch correspond to the width and thickness of the wall plates. When the roof goes up, the rafters are arrayed so that their bird's-mouths sit flush onto the plates of the outside walls. (If you crave more terminology, the side of the bird's-mouth that seats horizontally onto the plate is called the "seat cut," and the side that's vertically flush with the outside of the plate is called the "heel cut.") The distance between the bird's-mouths and the true ends of the rafters is the amount of overhang. Like most other framing members, the rafters are spaced 16″ apart, and when they go onto the plates they're stacked directly above their corresponding studs.

The lower ends of the rafters are usually sawn off with two cuts, which makes them appear to come to a point in the manner of the pickets on a picket fence. The cuts are angled so that when the rafters are put up, one cut is perfectly horizontal and the other perfectly vertical. That makes it possible to nail a board along the vertical cuts of all the rafter ends and another board along the horizontal cuts, thereby sealing up the lower edges of the roof. The boards nailed along the vertical cuts at the lower ends of the rafters are called the "fascia," and the boards nailed up into the horizontal cuts are called the "soffits." So, if you didn't already, now you know what soffits are.

In most houses, sheets of $\frac{1}{2}″$ plywood are nailed down onto the rafters to become the roof decking. Then roofing felt—a paperlike substance saturated with tar—is usually nailed over the decking, and the final roofing material, generally shingles, is nailed over the felt. Needless to say, these procedures come with any number of exceptions. Although it's feasible to nail cedar roof shingles to an ordinary roof deck, it's preferable to dispense with plywood and instead to put up lath—narrow parallel boards with a few inches of space between them. Wooden shingles last longer if they can "breathe," and using lath instead of decking makes for better air circulation around the shingles. And regardless of what they're made of, shingles aren't the only kind of roofing. We'll discuss some labor-saving alternatives in due course.

We've arrived at the top of the house by now, but before going on, let's take one more look under the roof to observe a final structural detail. Since, as we've said, the roof is pitched, the rafters bear down *angularly* on the exterior walls that support them. Not only does their weight try to push the walls vertically downward, it also tries to push them horizontally apart. Something, therefore, had better counteract that outward thrust by pulling the tops of the walls together. If the house has an attic, the problem's solved. The attic floor joists, being nailed down into the plates and also into the sides of the rafters, are more than sufficient to counteract the outward thrust of the rafters. But if there's no attic—I'm bringing this up because our outbuilding won't have an attic—and the ceiling under the roof is flush with the bottoms of the rafters, something else has to hold the tops of the exterior walls together. *Ergo* tie beams! Tie beams are beams whose sole function is to counter a roof's outspreading thrust, and, since it takes tremendous force to grab both ends of a board and pull it apart, it's a rare roof for which generously spaced 2x4 tie beams aren't sufficient.

We've framed out our house now, but we still have to close it in. The first step is to nail sheathing onto the outsides of the exterior walls. Normally just $\frac{1}{2}''$ plywood, the sheathing is the basic envelope of the house, but it does a lot more than seal out weather. It also has structural importance. Geometrically speaking, a triangle is a very stable form. Unless you twist it out of the plane in which it lies, you can't easily change its shape. A rectangle, on the other hand, is inherently *un*stable; even within the plane it lies in, it can easily sway to either side and become a mere parallelogram. It can also collapse entirely, as any number of framing carpenters who've been lazy about nailing temporary diagonal braces to their unfinished structures have been rediscovering since the dawn of time. And because our frame house is composed just about entirely of rectangles, it needs something to keep the angles of its rectangles at an honest 90°. So, by rigidly holding the studs to the vertical, the sheathing braces the entire house. (Which is why, in actual practice, the walls of a house are usually sheathed before the roof members are installed; unsheathed walls are often too unstable to support a whole roof.)

Once the house is sheathed, all we need do is tack some builder's felt to the outside of the sheathing, nail whatever siding we choose over the felt (it can be clapboards, shingles, tongue-and-groove cedar, even vinyl or aluminum), stuff some insulation along the inside of the sheathing between the studs, put some more insulation either between the attic floor joists or under the roof decking between the rafters, run some pipes and wires through the walls and under the floors (drilling holes in studs and joists as we need to), nail some Sheetrock onto the insides of the studs and the undersides of the joists, tape and spackle and then paint the Sheetrock, nail some flooring to the subfloors (except for the bathrooms, where we observe the niceties and glue tile to the subfloors), nail some baseboards and other assorted trim around the rooms, and—*voilà!*—we have a house!

It's true we've omitted the odd detail here and there, like what to do about doors and windows and ventilating the roof and making stairs and banisters and closets and counters, but nonetheless we've just clicked through a fair number of the essentials of frame construction. It's simple, it works, it's accumulated an admirable track record, it's architecturally compatible with all but the strangest environments, and it's also what we'll use to build the outbuilding.

MATERIALS AND MODULARITY

Things aren't always what the nomenclature says they are. A 2x4 isn't 2″ thick or 4″ wide; it's only $1\frac{1}{2}''$ thick and $3\frac{1}{2}''$ wide. Early in its evolution, when it was first rough-cut timber, it might have been a 2x4 that measured 2″ by 4″. But the board surrendered moisture through evaporation and consequently shrank, and then it was milled to smooth its surfaces and bevel its edges, and along the way it lost $\frac{1}{2}''$ in each dimension.

This is something that all professional carpenters know at least as well as their own names, and it qualifies them to patronize people who *don't* know it. When a carpenter's marriage fails and he indignantly recites his wife's offenses to a divorce lawyer, the lawyer may take on a superior smile and say, "C'mon, now. You don't expect to get anything like justice in court, do you?" For the moment the carpenter may feel put in his place, but later, when they're talking about construction (and they

Boards stacked at a lumberyard

will; it's inevitable), the carpenter may create the chance to say, "Hey, you didn't think a 2x4 is really 2″ thick and 4″ wide, did you?" Now it's the lawyer who's been put in his place. Parity has been achieved, the carpenter and lawyer are on something like an equal footing, and the social order has accordingly become more stable.

However, unlike the shifty 2x4, a standard 4x8 sheet of plywood is actually 4′ wide and 8′ long, a fact that has important implications for our outbuilding. We talked about studs and joists being spaced 16″ apart. Putting that into terminology, we'll say that the joists and studs are "on 16″ centers," i.e., that the distance between imaginary parallel lines drawn through the centers of the 2″ sides of the joists or studs would be 16″. Well, imagine a rectangle whose sides are two adjacent wall studs, whose bottom is the sill on which the studs are standing, and whose top is the plate across the tops of the studs. That rectangle is called a "bay." Its height is around 8′ (we'll be specific in due course) and its width is $14\frac{1}{2}″$. (It would be 16″ if studs had a thickness of zero, but since they're $1\frac{1}{2}″$ thick there's that much wood to account for.) A stud wall is made up of a series of such bays, and it won't take you long to realize that if you nail a 4x8 sheet of plywood to a stud wall, you can do it so both its long edges fall on the centers of studs. You'd have $\frac{3}{4}″$ on each of the outer studs (half their $1\frac{1}{2}″$ thickness) to nail the edges of the plywood to, and you'd be left with another $\frac{3}{4}″$ on each stud to nail adjoining sheets to the left and to the right. The sheet of plywood would cover exactly three bays.

Now, when you build the outbuilding you want to minimize both cost and labor. Once you've bought plywood floor decking or roof decking or siding or Sheetrock, you want to use whole sheets wherever you can. Besides reducing waste, using whole sheets lets you take advantage of their smooth, machine-cut edges.

Consider the deck of a small one-story building. Let's say the building is 16′ long and 12′ wide, and it's time to nail down the plywood subfloor. In subfloors, the plywood sheets are generally laid so that their long edges are perpendicular to the joists, and they're always set down in the same alternating pattern in which bricks are usually laid. Just as with a brick wall, alternating the elements breaks up the seams and gives you more strength. Yet, even holding to the alternating-pattern rule, installing a 16′ by 12′ subfloor requires only a single cut in a single sheet of plywood. If you divide the floor into three 16′ strips, each one 4′ wide, two whole 4x8 sheets set longitudinally cover the first and third strips, and one whole sheet and two half sheets cover the middle strip. Moreover, if you let the edges of the half sheets that you've cut yourself face the outside of the building (where, maybe thankfully, they'll eventually be covered by siding), all the edges of all the plywood inside the subfloor will be machine-cut, which means they'll all butt against other machine-cut edges and give you nice smooth seams. And since, as we've already seen, the 16″ spacing of the joists enables the 4′ edges of every whole sheet of plywood to fall on the center of a joist, you can nail those edges down to the joists just as good practice dictates. The resulting subfloor will look astonishingly professional.

Benefits like these practically mandate that our outbuilding be dimensioned in 4′ modules. In fact, the outbuilding whose construction we'll follow in forthcoming chapters is 16′ long and 12′ wide. We'll talk about varying those dimensions, but all the variations we discuss will be in units of 4′. Even within those constraints, you'll more than satisfy whatever cravings you have for measuring, sawing, and calculating. Outside those constraints you risk the possibility of massive overdose.

SITING AND RED TAPE

Sometimes, despite the everyday profusion of hard and painful lessons, ease begets ease. In this case, making the outbuilding easy to build spins off the ancillary benefit of making it easy to choose a site.

Aesthetics aside, the rules of thumb that govern site selection for an ordinary house are largely aimed at keeping its basement dry. Hence the usual injunction to stick to high ground. The traditional bugbear of basements is runoff—rainwater that flows downhill until it encounters your foundation walls and eventually seeps in. You're therefore commonly enjoined to site your house so that water flows away from it, not toward it.

But concrete foundations are expensive to buy. You can make one yourself, but the work is tedious and difficult, and in the end they're still expensive. Which is why I'm recommending that you build the outbuilding on pressure-treated wooden posts. Not high ones, mind you: the house sills won't be any higher than if they were set on a poured-concrete or concrete-block crawl space. But setting the building on posts gives you immensely more latitude in choosing a site. If you have some favored hillside that flows with sheets of water during downpours, use it anyway. Water will pass freely under the outbuilding and leave it high and dry. The only really relevant consideration is mud, and even then it applies more to you than to the building. Unless you're building a gazebo for your hippopotamus or an aviary for mosquitoes, pick a place that you can walk to without needing frequent recourse to your local quicksand rescue squad.

Another factor worth considering is light. If you like it, site the building in open terrain and have one of its long sides face southward toward the sun—being long, it can accommodate more windows. If you don't care for light, choose the north slope of a densely forested hill.

Moving from the realm of the obvious, though, it's important to discover what your legal setbacks are. Different local governments, usually villages or townships, often have zoning ordinances that regulate how close to the boundaries of your land you're allowed to build. Sometimes there's a setback formula that takes into account the height and nature of the building and even the slope of the site. The most cogent thing I can say about investigating what zoning restrictions govern your intended outbuilding is that early is probably better than late. It's dispiriting to put up a whole building only to be told that you have to move it or tear it down.

And while you're visiting your building department, you might also find out if you need a building permit. Most local governments will let you erect an outbuilding fairly readily; what worries them is that you'll put up a whole other house. So prepare to be forbidden to install any plumbing. To building departments it's generally kitchens and bathrooms that define a dwelling, and if your outbuilding is classified as a residential structure you'll either be told to forget it or you'll have to hew to a much more stringent set of building-code requirements than would apply to a structure that's *not* residential. A lot of this depends on how rural your property is. As a general rule, the closer you are to civilization the larger the presence of the building inspector looms. In the bona fide boondocks he's barely a presence at all; in the suburbs he's a snoop and a stool pigeon. The great value of calling your intended structure a shed or an outbuilding is that your building department is likely to be even more casual about it than they'd be about a garage. However posh their insides are, outbuildings are normally considered to be commodious toolsheds, and getting your building department to see it that way will be all to the good.

One final caveat. In line with Buber's notion of imagining the real, spend some time imagining how your building materials will get to the site. If the lumberyard's delivery truck can't come in nearby, you'll have to move some dauntingly unwieldy sheets and boards. I won't recommend that you site the building near an accessible roadway, because your heart's desire may be to have a studio that's *in*accessible. But if you put the outbuilding on, say, a steep, remote hillside, prepare to know in your bones what it was like to build the pyramids. Or at least prepare to buy, borrow, or otherwise enlist one of those giant plywood garden carts or a power winch or a squadron of sturdy friends or some combination thereof. In the end it's better to site the outbuilding where you want it and not where it's easiest to build. But privacy will have its price. If it's any help, try to remember that there were epochs in which exercise was practiced not on Nautilus machines but on materials that had to be moved from one place to another.

2

Making a Sketch

Experimenting on paper is less cumbersome than experimenting on lumber, and, what with all the many factors that will influence the project, the best time to account for them is at the planning stage. Clearly your sketch will depend to an extent on your intended site, on how you plan to use the outbuilding, and on how your preferences in secondary shelter run. Since you'll be drawing windows and doors, it will also depend on what materials you plan to use. Those will vary with both budget and taste, and two more mercurial concepts may not exist. On a per-square-foot basis, doors and windows are likely to be the most expensive materials you'll buy, and their prices range much more widely from brand to brand and vendor to vendor than the price of ordinary lumber. Real bargains can be had, but you've got to shop for them, and, predictably enough, they generally come at the expense of quality. Then again, you probably won't want to impose the same quality standards on an outbuilding that you'd impose on a main dwelling. Outbuildings have an aesthetic all their own, and it's appropriate that even new ones look a touch *trouvé*, in the manner of restored antiques.

So, necessarily, your drawing will be a many-layered entity. If its manifest subject is the plan for the outbuilding, its subtexts are your taste, your budget, perhaps your willingness to scavenge at the dump, perhaps your luck at finding windows at whatever your local version of Salvage City is, and certainly your ability to keep your ambitions for the project in line with what you can actually accomplish and afford.

And if all that makes the prospect of doing the drawing seem forbidding, a pair of mitigating factors brings it down to size. The first is that in most cases the drawing need only be a sketch and not a blueprint. Whereas building departments want to see signed and stamped architect's plans before they issue a building permit for a house, they seldom require more than a loose rendering for an outbuilding.

The second factor is that you can alter the drawing whenever you like. In the end it's really just a memorandum to the builder (in this case you), not a contract with destiny. Actually, so great is the likelihood that you'll change the design as you go along that for now the only part of the drawing we'll go into in any real detail is the deck. Sketch the rest—walls, roof, doors, and windows—however you like. And when you do sketch them, try to arrange with yourself not to become inextricably attached to what you've drawn. Putting up the deck (with all the deeply edifying trips to the lumberyard that that entails) will almost undoubtedly alter your perceptions of the building, of the materials that go into it, of the money you want to spend, and even of yourself as a builder-designer.

By the time you've built the deck you'll have learned a whole lot through sheer experiential osmosis, and, just as important, you'll be able to walk the floor of the outbuilding for real, not just in your imagination. Lines of sight through intended windows will come clearer, and so will the way ingress and egress are affected by the surrounding landscape. You'll also have had some time to browse the door and window catalogues at the lumberyard and perhaps look around for castaway doors and crippled windows. And once you've built the deck you'll begin to *see* how a battered but interesting-looking old door can be restored or how a window that's been left for dead might be interestingly rehabilitated. If I use the word "imagination" often, it's because building anything substantial has a way of conditioning your imagination, and after the deck is done you'll almost certainly want to adjust your plans for entryways and fenestration and maybe other things as well. In fact, I'll go so far as to say that you *should* change your plans at least a little. To refrain from taking advantage of everything you're learning would be downright foolish.

If you don't have much money to spend and your local dump is a forager's paradise, it may also be foolish not to take advantage of free materials. Poking around at the dump is definitely one of the more dignified forms of scavenging, even if it means you have to redesign a bit to accommodate what you find.

Determining the Size of the Outbuilding

To begin getting your design down on paper you'll have to decide how big your outbuilding will actually be. Again, the one whose construction is illustrated in this book is 16' long and 12' wide, but yours can be bigger or smaller. In line with what we've said about modularity, I strongly suggest that you choose dimensions which are multiples of 4'. And if the availability of helpers is likely to be sporadic (even the best-meaning people tend to become listless and unfocused when working on someone else's project), you'll be wise to keep the outbuilding on the small side. Handling big boards begins to get hard if there's no one around to grab the other end. If you use the deck as a giant work table to frame your walls on, you'll find that a 16' wall frame is about as long a one as you can raise yourself. (Notice that I didn't say "comfortably raise" and I certainly didn't say "safely raise." Standing up a 16' array of 2x4s without any help may be possible, but under no circumstances can it be said to be comfortable, and it definitely isn't safe.)

Building wall frames on the ground—actually the deck—and then standing them up after they're put together is sometimes called "Western-style" framing. For the novice it's probably the easiest way to go about things, but it isn't the only way. If you're troubled about the prospect of nailing gigantic wall frames together only to find that you can't raise them up, you can always assemble them vertically in place and then not have to move them at all. (That approach may or may not be called "Eastern-style"; construction lexicography is a fuzzy field.) Building walls in place is a little more demanding of your technique with a hammer, in that you have to do more toenailing—driving nails at an angle to the board—as opposed to the more familiar face-nailing. But it's standard practice with a lot of carpenters, and it's certainly an available alternative. Besides, you'll have to do a fair amount of toenailing anyway, and the more you do the easier it gets.

Of all the factors that constrain the size of the outbuilding, though, the one that most merits serious consideration is the difficulty of putting up long roof rafters. Especially if you're inexperienced and working by yourself, it simply doesn't feel good to be up on a ladder trying to set into place and then nail rafters—even the relatively light 2x6s we'll be working with—across a horizontal span much greater than 6'. And since the horizontal span between the centered ridgepole and the side walls of a 12'-wide outbuilding is 6', I'd suggest limiting the width of your structure to 12' unless you're pretty sure you'll have a strong, patient, competent, empathic collaborator around when you're framing the roof. If you *are* sure, then you can safely go to 16'. Past that, building codes would require you to use 2x8 rafters, and long 2x8s are cumbersome devils to work with.

The remaining limits on the building's size are really structural ones imposed by the basic deck design. The deck, and thus the entire outbuilding, will rest on three parallel 6x6 girders, each running the full length of the building. There'll be one girder under each of the long outer edges and a third girder midway between. Each of the three girders will in turn be supported by three 6x6 posts, one at the center of the girder and one near each of its edges, making a total of nine posts for the building to rest on. Running at right angles across the tops of the girders will be 2x6 floor joists on 16" centers. So, if you take a moment to think about it, you'll see that the limits on the size of the building are functions of (1) the maximum permissible span between posts along a 6x6 girder and (2) the maximum permissible span along 2x6 floor joists between girders. Observing those maxima ensures that the girders don't sag too much between the posts and that the floor doesn't sag between the girders. Current standards hold that, for normal floor loads, the maximum girder span between posts is about 10', while the maximum span along 2x6 floor joists on 16" centers is about 8'. (What I mean by "about" is that the actual maxima vary somewhat with type of lumber; the spans I'm giving you should be safe for virtually any construction-grade lumber you can buy.) Doing some simple arithmetic, we can conclude that the practical maximum size of the outbuilding is 20' by 16'.

Roughly Estimating Cost

Now, knowing what you know, deliberate about the factors we've discussed. Think about the number of trees you're willing to cut, the amount of lumber you're realistically prepared to move to the site, the longest 2x6 rafters you're willing to juggle around up on a ladder (as a crude rule of thumb, figure the rafters to be a third longer than the horizontal distance they'll span, making the rafters for a 12'-wide outbuilding, which have to span 6' horizontally, around 8' long), and of course how much money you're ready to spend on lumber, roofing, siding, and the like. On this last point I'll give you another crude rule of thumb. Excluding windows and doors, figure the lumber bill for a 12' by 16' outbuilding at $1,000, although, depending on where you live, that amount may vary by as much as 50% or even more. Proximity to trees and mills, cost of transportation and warehousing space, and the vagaries of the lumber market all play so significant a role in your actual bill that there's little point in my telling you what it's likely to come to. But calculate the square footage of the outbuilding you want to build by multiplying its length by its width, then compare it to 192 square feet (12 times 16). In fact, obtain a percentage by dividing *your* square footage by 192 and then multiplying by 100. Now take that percentage of $1,000 and you'll have the barest rudimentary inkling of how much your lumber will cost.

Starting Your Sketch

Finally, having brooded upon all those issues and digested them as best you can, pick a length and width that are both multiples of 4' and draw a rectangle proportioned more or less accordingly. You now have the outlines of your drawing.

(Incidentally, the terms "length" and "width" can become confusing if you're putting up a square building, which is a perfectly reasonable thing to do. To avoid confusion, therefore, we'll adopt the following convention: for a square building, the ridgepole atop the roof and the girders beneath the building run along the *length* of the structure, while the floor joists run along its *width*.)

Now proceed with your sketch by superimposing an aerial view of the three girders onto your rectangle. Don't worry about sketching exactly to scale quite yet (if ever), but from the size of your rectangle and the dimensions you've assigned to the deck you can eyeball the equivalent length of a foot in your sketch, and if you take half of that you'll have an approximation of the thickness of the girders (they being 6x6s). Draw one girder going the long way down the middle of the deck and the other two, parallel to the first, along each of the edges. The girders run the full length of the building, but the sides of the outer girders don't have to come out to the edges of the subfloor. You can safely set them in up to a foot from the long edges of your proposed floor without incurring structural problems. The floor joists will simply cantilever out over the girders to support the extra foot of floor, and then the outsides of the girders will ultimately be less visible to the outbuilding's envious beholders. It's a look you may or may not want, and the choice rests entirely with you.

Now that you've sketched the girders, mark their centers and superimpose a little square, half a foot on each side, where the center posts will support them. Then do the same for the tops of the outer six posts. If you like, the outer posts can also be inset up to a foot from the ends of the girders. This time it will be the girders that do the cantilevering; 6x6 girders can comfortably cantilever one foot of outbuilding. It's really a question of how much post you want to expose to the outside world. A foot of cantilevering is one look, six inches of cantilevering is another look, three inches of cantilevering is, curiously enough, a quite different look, and *no* cantilevering is still a different look. What's more, I offer no suggestions. Like matters of budget, labor, and time, questions of aesthetics are ones in which this man's piddling nuance is the next man's gross disparity. You're on your own.

Having chosen where the posts will go, you're ready to draw the floor joists. The joists are nailed across the tops of the girders and in your drawing will run along the full width of the building on 16″ centers—almost. Your 2x6 joists are, as we said, $1\frac{1}{2}″$ thick, and if you obey the 16″-centers rule to the letter your bays will have exactly $14\frac{1}{2}″$ between each joist. Assuming the length of your outbuilding is a multiple of 4′, you'll find when you start drawing the joists that the last one just falls off the edge of the girders. (The distance across four 2-by-anythings on 16″ centers is $49\frac{1}{2}″$, and 4′ stubbornly remains 48″.) Therefore, in order to squeeze the last joist back on, you'll need an extra $1\frac{1}{2}″$ of girder, which you'll get *not by lengthening the girders but by shortening two of the bays*. Why two? You want all the edges of your sheets of plywood decking that run parallel to the joists to fall on solid wood so you'll have something to nail them to. And if you shorten the two outer bays—i.e., the bays at either end of the subfloor—by $\frac{3}{4}″$, everything will tumble nicely into place. Any distance that's an exact multiple of 4′ will, if measured along the length of the building from the edge of the subfloor, fall on the *center* of a joist, which means that the seams between your factory-cut subfloor sheets will likewise fall on the centers of joists, allowing you to nail their butted edges to the same joist. Moreover, when the deck is long done and you're putting up plywood siding, you'll reap the same advantage all over again. Since your 2x4 wall studs will stack directly over your joists and therefore be spaced the same way, the factory-cut edges of the siding will similarly fall on the centers of studs, allowing you to nail *their* butted edges—well, actually the edges of plywood siding are lapped and not butted, but don't concern yourself with that for now—to the same stud. And all you need to do to make that happen is shave $\frac{3}{4}''$ off the bay at either end of the deck, so that the space between the joists is $13\frac{3}{4}''$ instead of the usual $14\frac{1}{2}''$.

That done, draw a board of indeterminate thickness running the full length of the building along each side. This pair of boards will box in the ends of the joists, and although it's simplest to make them 2x6s like the joists themselves, they can also be lengths of 1x6 shelving. You don't have to decide what they are right now, but make a note to remind yourself to cut the joists so that the total distance across the deck, *boxing included*, equals the multiple of 4' that you've settled on. (If you're grumbling about having to cut joists at all, perhaps because you've decided to make your outbuilding 12' wide and happen to know that your lumberyard stocks the very 12' 2x6s you'll be using, you can stop grumbling. Twelve-footers can usually be relied on to be *at least* 12' long, but rarely are any two exactly the same length. You'll have to trim them in any event.)

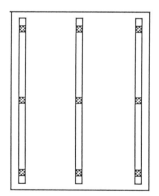 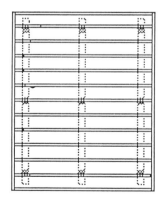 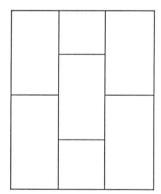

Complete the overhead drawing of the deck by sketching in the plywood subfloor. Place the 4x8 sheets so that their long sides run *across* the joists. Even plywood has grain, and, like most wood, it bends less easily along its grain than across it. As we mentioned before, don't forget to alternate the sheets in a brickwork pattern. And make sure your pattern lets you use as many whole sheets as possible.

 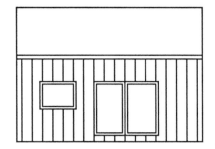

Now that the deck is a known quantity, make a few simple elevations of the sides of the building too. (Elevations are full-on side views.) You're still just imagining, not forging a commitment. Keep the roofline simple and symmetrical, don't put in too many more windows than you can afford, and, since small buildings tend to be annoyingly short of wall space for shelves, furniture, coat hangers, and the like, try to make do with only one door. Be generous with yourself, but be circumspect too. Even if you can afford to treat yourself lavishly, don't do it at the expense of practicality. Lots of windows are fine, but people who live in glass outbuildings frequently have no place to hang a picture. And speaking of wall space, if you plan to put in a woodstove or a wall-mounted gas heater, work out how much wall space you'll need. It's disheartening to trim out a window with a flourish of craftsmanship only to watch your woodstove slowly—or suddenly—turn your trim to charcoal. Generally speaking, this is the time—while there are still no nails to pull or framing members to pry apart—to assess your needs and provide for them in your sketch.

Then, when you've finished drawing, take what you have to your local building department. If they laugh in your face, see if you can penetrate the mirth long enough to find out what you have to do to be in compliance, and redraw or resite so you can try again. If they tell you it's acceptable and either give you a building permit or tell you you don't need one, go to your proposed site, stoke up your imagination, and plunk the proposed deck down onto it. Surround yourself with the proposed walls and live for a few moments in the proposed structure. If you think you want to *keep* living in it, go on to the next chapter.

3

The Foundation

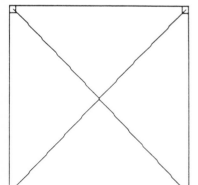

Laying Lines

Making sketches is well and good, but sooner or later, imagining the real has to give way to *realizing* the real, and, unless it's winter and the ground is frozen hard, that time may as well be now. So go to your site, measure off the dimensions of your outbuilding, bang in a stake at each of the four corners (straight pieces of tree branch are as good as milled wood, and a rock is as good as a hammer), and connect the stakes with twine. (The best twine for this purpose, readily available at most hardware stores, is yellow mason's twine, mainly because it's yellow and stays yellow. Twine that's already, or is likely to become, the color of your site will soon be invisible, at which point it will try to trip you or girdle your ankles and stanch the supply of blood to your toes.) You'll discover that there are practical limits to how much precision the marking-off process will bear, so don't tie yourself in knots over fractions of an inch. The main thing to strive for is a true rectangle, and the best way to ensure that its angles are all *right* angles is to keep both diagonals equal. Which brings up the fact that one tool you'll have to buy or borrow for the project is a measuring tape that's at least as long as a diagonal of the outbuilding. For geometry fans, that's the square root of the sum of the squares of the length and the width, a dimension easily obtained with a pocket calculator.

But another thing not to tie yourself in knots over is calculations, at least not now, when you're working with something as imprecise as earth. Regardless of what your calculated diagonals are, in the field just get them equal. True right angles are more important here than an inch or two of length or width.

Clearing Brush and Trees

Since you're building on posts, you won't have to do any grading, but tradition dictates that the next step after marking off the corners of the building is clearing the land. So cut down any trees inside your rectangle and keep the stumps shorter than the bottom of the building. If there's a stump where any of your posts will go, remember that right now it's much easier to move the building a few inches than to pull a stump. And give some attention to the trees just outside the rectangle. Trees in close proximity to a new structure tend to make it look less raw, less of an offense to the landscape. But follow them up with your eye. If any trunks or branches intrude on what will be the final volume of the outbuilding, get rid of them now. Once you've started with the carpentry, your momentum will be precious, and you won't want to interrupt it to become an arborist.

If you want to make any minor siting changes, this is also your last good opportunity to make them. Putting up even a small outbuilding is a strangely momentous venture. It doesn't always hit you at first, but eventually you realize that you're altering a part of the face of the earth. A measure of care and humility is warranted. Marry the building to the landscape as best you can, because presently you'll begin breaking ground.

Marking Out the Posts

Before getting repellently reverential, though, get inside your twine-bounded rectangle and bang in more stakes to mark off the centers of the nine posts. Start with the four nearest the corners of the building and use the dimensions from your sketch to locate their positions. (I'm assuming you've inset the posts at least a tad from the actual corners. If you haven't, placing the first four post markers will be all the easier.) Then run some

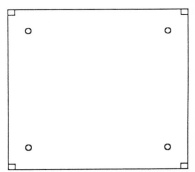

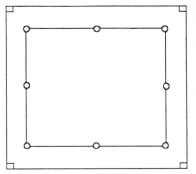

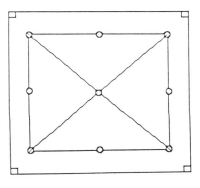

more mason's twine from each of the four post markers to the next, and you'll have a rectangle within a rectangle. If the sides of the inner rectangle aren't parallel to the sides of the outer rectangle, adjust them accordingly while keeping the corners square. But again, don't tie yourself in any knots. The bottoms of the postholes will be wide enough to let you do some fine-tuning before you carve your deck in pressure-treated wood.

Now find the centers of all four sides of the inner rectangle, mark them, and drive four more post stakes. You've now staked eight of the nine postholes. To do the ninth, just connect up the center stakes with more twine. The intersection of the two twine lines defines the location of the last, innermost post.

Digging Postholes

At this point you've driven thirteen stakes: four marking the corners of the building and nine marking the centers of the support posts. With an ordinary spade, cut a 2' circle in the ground around each of the nine post markers and start digging. Be forewarned that digging nine deep holes yields rather less personal fulfillment than just about any of the tasks you'll carry out in putting up the building. The work is boring, frustrating, and thankless, and it gets harder as the holes get deeper. Granted, if you've sinned against mankind recently and need some self-mortification, digging nine deep holes may be just the ticket. But if you've behaved decently, reward yourself by calling local hardware stores or equipment rental services and seeing if you can rent a gas-powered posthole digger. Even if you're working in the most compliant, rock-free soil, once you're down a couple of feet you won't be able to level your shovel inside the hole, which means you'll be removing dirt a teaspoon at a time. Mechanical posthole diggers aren't nearly as formidable as they look, and if you rent one you may be able to accomplish in a day what might otherwise take you weeks.

Oh, yes. The depth of the holes. Make them 3' 8" deep—3' for the post and another 8" for the thickness of the poured concrete. That way you'll be protected against frost heave in all parts of the country, and you won't have to brace the posts to keep them from leaning away from the vertical. When you backfill the holes after you put the posts in, the ground will provide all the bracing you need.

Digging postholes the hard way

With the digging done, you're ready now to pour your footings, those being the concrete pads that keep the bottoms of the posts from sinking into the ground beneath them.

If you've never fooled around with concrete or mortar before, there's no reason to tense up. The stuff can be bulky and sloppy, but it won't ask for any special skills. It's true that when an application calls for concrete to be inordinately hard or free of porousness a certain degree of knowledge and experience is warranted. But all you require of your footings is that they remain more or less in one piece and not compress under the weight of the posts.

About Concrete

Just so you know what you're working with, everyday concrete is a mixture of Portland cement, sand, and small stones and/or crushed rock. The last ingredient is called "aggregate." While sand and aggregate are found in nature, Portland cement is manufactured. Limestone is pulverized with shale, clay, or marl and then heated to about 2,700° F. After some time the mix fuses into a "clinker," which in turn is ground into the fine powder that's sold as Portland cement. Different grades of cement are obtained by varying the ratio of limestone to clay, but the two most commonly available grades are mortar-type cement, which you mix with clean sand (around three parts sand to one part cement for a good strong mix) to make brick mortar, and concrete-type cement, which, as we said, is mixed with sand and aggregate to make concrete. The proportions of cement, sand, and aggregate vary fairly widely from one concrete mix to the next, but a typical, middle-of-the-road mix might contain three parts sand and four parts aggregate to one part cement.

Curing Concrete

As for how cement hardens, it's got nothing whatever to do with "drying." In fact, Portland cement hardens just as well underwater as it does aboveground, and sometimes even better—a property that greatly facilitates the building of bridges and tunnels. While the various ingredients of Portland cement, most of them compounds of silicon, lie around in the ground for all those many millennia, wondering when they'll be mined and dumped into cement kilns, they combine chemically with water. Then, in the high temperatures of the kilns, the molecules of water are driven off as steam. However, the ability of those compounds to combine with water once again is unimpaired. So when you add water to Portland cement, they do combine with it—chemically—and the resulting crystalline mass is what gives concrete its stonelike strength. In fact, concrete permitted to dry out while setting isn't as strong as concrete that's been kept damp until it hardens. But unless you're putting up your outbuilding in the desert, the ground at the bottom of your

Mixing concrete

postholes will almost certainly not be dry, so your footings will enjoy the benefit of damp-curing without your having to take any particular pains one way or the other. And if you *are* building in the desert, you might wet the stuff down a little once or twice a day for a couple of days—just enough so that it never completely dries out.

We'll do some very rough calculations now. Figure the bottom of each of your postholes to be a circle about 1′ in diameter, and figure their thickness to be around 8″. That would make the volume of each of your footings a shade over half a cubic foot. (In feet, $\pi r^2 h = 3.14 \times (\frac{1}{2})^2 \times \frac{2}{3} = 0.5233$ ft³.) Which means you'll need around five cubic feet of concrete to pour all nine footings.

Premixed Concrete (Add Water and Stir)

If you don't have any Portland cement around, the easiest way to do the footings is simply to go to your hardware store or local brickyard and buy nine 80-pound sacks of premixed concrete (Portland cement, sand, and aggregate are already sifted together; Sakrete is the best-known brand but hardly the only one) and allot one bag per footing. Each 80 pounds makes a shade over half a cubic foot, and all you need to do is follow the directions and add water, mix until fairly uniform (if you don't have an old wheelbarrow for this purpose, you can use a garbage can or even one of those plastic washtubs), shape the dirt at the bottom of each hole so it forms a circle that's roughly a foot in diameter, tamp down the bottom of the hole with the end of a board so the dirt is fairly compressed, then pour in the wet concrete. Make sure to add sufficient water so that the top surface of the concrete levels more or less by itself. You can encourage additional leveling by slopping the end of a stick in the wet mix. And while you're doing that, look for any chunks of aggregate that prominently protrude above the surface and push them back down. When the mix hardens you'll want a reasonably smooth surface on which to move the bottoms of the posts around as you make the final adjustments to their position. But also keep in mind that all you're making here is a buried footing, not a skating rink. A modicum of fussing is appropriate, but don't tip into compulsivity.

Mixing Your Own Concrete

If you do happen to have some Portland cement around (or if you prefer to mix your own concrete just as an exercise), a mix of one part concrete-type Portland cement to three parts clean sand and four parts small stones should do fine. Proportioned at that rate, a single 94-pound sack of Portland cement should make about 8 or 9 cubic feet of concrete, which is more than enough for the whole job. Portland cement is available by the sack at brickyards and masonry supply houses and even at a fair number of hardware stores, but unless you already have some sand and aggregate near your building site I wouldn't recommend going to the trouble. Even if you do have the sand and aggregate you'll save only a few dollars by making concrete from scratch. Most masonry suppliers won't deliver sand in quantities of less than a cubic yard (that's 27 cubic feet). As for aggregate, I assume you'll be using your own stones, which you really ought to pile up and hose clean before you mix them into concrete. Still, if you're adamant about making concrete, mix the dry ingredients until they're fairly uniform, then add water a little at a time and use a hoe (or something hoelike) to draw them into the puddles of water. Keep doing that until everything is wet, then—still only a little at a time—add water until you can move your hoe fairly easily through the whole mix. When it has the consistency of fresh oatmeal you're ready to pour it into your postholes, their bottoms shaped as described above. Also as described above, do what you can to encourage the concrete to make a level surface without any prominent bumps. Home-grown aggregate tends to be more chunky than the store-bought kind, so try to push the bobbing heads of any stones down into the wet concrete, however cruel it makes you feel.

Your footings will need at least a day—maybe two or three days; it depends on the outside temperature—to get hard enough to work with, so what's wanted now is some measure to prevent your postholes from caving in onto the footings. The chief danger here is a severe rainstorm, not because the concrete will be harmed by the water but because the rain may cave in the walls of the holes and bury your footings prematurely. So look around (the dump is a good place) for scraps of plywood or even heavy cardboard. Lacking those, make some mini-tarpaulins out of clear plastic—10x25 rolls of 4-mil plastic sheeting are often sold in hardware stores for only a few dollars—and secure them

Concrete going into posthole for footing

over the holes. You can use stones to hold them down, or you can peg them right into the earth with sharp sticks. However you handle the problem, try not to *be* the problem. Which is to say, don't accidentally cave in the dirt around the sides of the holes while you're trying to prevent the dirt around the sides of the holes from caving in.

Concrete Piers

Your footings are setting now, you're thinking ahead to the next step, and somehow you find yourself harboring the wish to have the building rest on posts made of some more durable material than wood, albeit pressure-treated wood. Well, just so you can't say you weren't alerted to the possibility, you can always go to your local masonry supplier, buy cardboard tubes manufactured to serve as concrete forms, and pour concrete posts to your taste. You'll have to mix a lot of concrete or have some ready-mix delivered, and you'll also have to be much more careful with placement of the forms. Once they're full of concrete they're not going to move around as easily as pressure-treated 6x6s. In fact they'll *never* move, and any mistakes you make will literally be etched in stone. And I strongly suspect that if you opt for concrete posts you'll have barely begun making them when your wish for archival permanence will evaporate abruptly. Pouring nine of them by yourself is a messy, expensive, time-consuming, and generally unattractive job. The great thing about pressure-treated posts is that you cut them, drop them into the holes, line them up and get them more or less plumb, and there they are, ready to hold up a building. Besides, all over America insects and microorganisms are sampling chunks of pressure-treated wood and going "phooey" if they're lucky, dying if they're not. And if the posts do rot out over the course of time (probably a good couple of generations), so what? Imagine how delighted your heirs will be at the prospect of a juicy restoration project. Meanwhile the epic struggle between modern chemistry and the ancient agents of decay will be playing itself out right under your feet, and the outbuilding will seem to crackle with its energy. Go with it.

4

The Deck

ABOUT ORDERING MATERIALS IN GENERAL

So far the only materials you've had to buy are cement, perhaps a pad of paper, and a long tape measure. But to go on and build the deck you'll need more than the odd stick of lumber, and now you're faced with the notion of spending more than the odd dollar on a project you may still not be entirely certain you can carry off. I could reassure you here by telling you it's all more manageable than, in your more pessimistic assessments, it might appear. For one thing, you'll be working on the ground, which is easy to get to. For another, you'll be using lumber that, because of both its size and final destination (i.e., under the building where it won't be seen), can soak up quite a few mistakes and not show many of the scars. But if you're nagged by second thoughts, I'm not convinced it's reassurance that's appropriate. The better idea may be to realize that there's risk attendant to the start of any sizeable endeavor. Yellow fever nearly kept the Panama Canal from getting past the mudhole stage. You, having just poured your footings, are *at* the mudhole stage. So yes, you're about to take a plunge, but why not let the thrill of risk make the outset of the project all the more bracing?

Of course, the easiest risks to live with are calculated ones, and one way to infuse your risks with a comfortable degree of calculation is to purchase your materials wisely. If you hate shopping, a wise purchase is a fast purchase, and the fastest way to buy lumber is simply to open a charge account at a good all-purpose lumberyard. Even if you've done time in debtors' prison, you shouldn't have a problem getting a lumberyard charge account. Professional builders make up the bulk of the account holders at most lumberyards, and builders are notoriously remiss about paying their bills. Which means that lumberyards are usually more than happy to start an account for a layman. However good you think you are at ducking creditors, it's the rare builder who couldn't take you through a master class in the fine art of invoice management. Lumberyards know this very well, and they also know that having an account with them is going to discourage you from shopping around. Once you're deep in mid-project you tend to be the prisoner of your own momentum: you hate to squander it by pausing to compare prices. And sometimes comparing prices *is* a waste of time. If

there are several lumberyards around, ordinary free-market pressures will tend to keep their prices pretty much in parity.

Far more important, I think, is that you arrange to get a builder's discount. An outbuilding's worth of materials should be sufficient quantity to get it, and if one lumberyard doesn't give it to you, try another. Builder's discounts are typically structured so that you get a flat 10% or so off every bill and another 5% or so if you pay before the next month's bill comes. And be wary of lumberyards that charge for delivery. A 16'-long pressure-treated 6x6 (heavier than not pressure-treated) isn't conveniently transportable in the family car, and no one expects you to own a truck. For builders, delivery costs are already factored into the price. They should be factored into your price too.

One last thing: order materials on an as-needed basis. There's little point in storing (and paying for) lumber that you won't be using for some time to come. Your lumberyard is unlikely to reward you for taking early delivery by giving you a rebate for warehousing, and having piles of extraneous lumber on your property is bound to make for clutter and confusion. But don't order too little at a time either. It's jarring to run out of materials just when the fires of carpentry are burning brightest; sometimes you're left so high on tenterhooks that you can't do anything else until your next delivery arrives. A good rule of thumb is to have each order contain enough lumber to get you through a chapter. In this chapter we'll go over what you'll need to build the deck.

MATERIALS FOR THE DECK

Here, from the ground up, is what you'll need:

1. posts
2. girders
3. truss plates
4. nails
5. joists

6. boxing
7. Styrofoam (optional, but recommended)
8. plastic vapor barrier
9. plywood decking
10. shingle scraps for shimming
11. scrap lumber for later

Let's take them an item at a time.

1. Posts

After all nine of your posts are resting on their footings and standing more or less straight up, their tops should eventually describe a level plane regardless of the depth of the postholes or the topography of your site. The deck of the outbuilding will rest on the tops of the posts, and you want your building to be level. You'll start out with posts that are somewhat larger than they need to be and cut them off, but for now you want to know how much wood to buy. There are endless ways to do the reckoning, so I'll suggest only one. If you prefer another way, use yours. Field expedience demands that you do whatever suits you best.

You'll need a line level now (unless you're handy with a transit), so borrow or buy one (they're cheap). Being careful not to cave in any dirt, drive a stake into the ground just beside your highest posthole (i.e., the one on highest ground) and another, longer stake into the ground next to your lowest. Make the stakes sturdy enough and drive them in far enough so that you can connect them with twine that's fairly taut without moving or breaking them. Tie the twine first to the higher stake at a point just above ground level. Then, before you tie anything else, imagine the twine proceeding to the other, lower stake along a level line. If you find yourself having trouble imagining a level line (some topographies defeat all attempts to guess at what a level line is, even to the point of making truly level houses appear tilted), go ahead and temporarily tie the other end of the twine to the lower stake and play with it until it's level. Keeping the twine taut, measure off the halfway point between stakes and hang your line level from that point on the twine. It shouldn't take more than a few tries to get the twine fairly level. Remember that the purpose of all this is to give you a feeling for where the tops of the posts will finally be. Girders will sit atop the posts, joists atop the girders, and flooring atop the joists, so the actual floor of your outbuilding will be another foot higher.

Now summoning your acutest powers of visualization, *see* the outbuilding at the level you've chosen and decide whether you like it there. Then, in your imagination, move it up or down a few inches and see if you like it any better. The tops of your posts needn't be more than about 4″ above grade (barring a deluge), especially since the floor framing will raise your floor well above that. And many people prefer small buildings to be low to the ground. On the other hand, leaving some room between the underside of the deck and the ground gives you covered storage space for firewood, screens, air conditioners (swaddled properly in plastic, of course), chain saws, axes, and spare lumber. You can always "bring down" the building visually by running lattice (the pressure-treated kind) along the bottom or planting shrubs around the base. A sloping site, if that's what you've chosen, will give you the best of both worlds: storage space under the "high" side and a nice low profile along the other.

When you've arrived at what you feel is a good altitude for the tops of the posts, use your level length of taut twine as a benchmark and measure down from your chosen altitude to the surface of the footing in your "high" posthole, then do it again for your "low" posthole. I was careful to suggest that you measure from your chosen altitude rather than from the twine itself, since you may have decided that you want the tops of your posts to be a little higher or lower than the twine and didn't feel like retying and releveling all that string. But regardless of the post height you've selected, you now have the approximate minimum length of your longest and shortest posts. As for the rest, you ought to be able to do them by eyeball. Unless your site undulates in an especially cockeyed way, create three categories of posts-to-be: those that are about as long as the longer of the two posts you've measured (i.e., the one destined for the hole in the lowest ground on your site), those about as long as the shorter of your two measured posts, and those that are halfway between. If your site *does* contain some abrupt changes of grade, you still ought to be able to estimate minimum post lengths from the two lengths that you've actually measured. This remains one of those areas in which too much precision is simply a waste of time. At all events, you'll buy enough wood to give you a substantial margin of error when you cut the posts, and if worst comes to worst you can always bring down the whole building an inch or two with little risk of having its floor be below grade.

When you're done reckoning, your posts will have come in somewhere around 4′. Now add half a foot or so to the length of every proposed post and call your lumberyard to find out in what lengths they carry pressure-treated, ground-contact-grade 6x6s. (Don't worry about the question marking you as an ignoramus; frame it in monosyllables, chuck in a few regionalisms, and the people at the lumberyard won't have any reason to doubt that you were born with a framing hammer in your little fist.) Common commercial lengths are 16′, 12′, 10′, 8′, and maybe 6′. Once you know what you can get, lay out your planned posts so you wind up with the least waste. And try as best you can to avoid using very long boards. They're heavy and therefore hard to handle. If you can make do with, say, two posts per 8′ board or three per 12′ board, you'll be happier than if you cut your posts from sixteen-footers. You won't save any money by using longer boards, either. The stuff is sold by the running foot, regardless of board length.

Now, presumably, you know how many pressure-treated, ground-contact-grade 6x6s, possibly of mixed lengths, to order for posts. Write down the numbers you've arrived at.

2. Girders

For girders, order three 6x6s, each one the full length of your outbuilding. If you can't get them that big (which could be the case if your building is longer than 16′), order six 6x6s, each one *half* the length of your outbuilding. When the time comes you'll simply butt the two halves together.

Incidentally, at about this point there arises the question of where to switch over from pressure-treated lumber to lumber of the ordinary variety. Almost universally these days, decks built on posts are framed *entirely* with pressure-treated members, but that doesn't mean you have to do it too. For the majority of small buildings, permanence is more a state of mind than a physical reality. All wood, pressure-treated or not, eventually rots or gets chewed by termites or carpenter ants; yet, throughout the northeastern United States, there are any number of wooden buildings dating back to the early 1600s, many of which are still inhabited and promise to go on being inhabited for centuries to come. Yes, the ones in use have probably seen greater or lesser amounts of restoration, but restoration is a perfectly normal part of building maintenance, and not just for wooden buildings either. Masonry crumbles too. Old bricks need to be repointed. Aging plaster walls need spackling and replastering. Foundations have a way of slowly disintegrating, regardless of what they're made of. Apart from sealing your outbuilding in a helium capsule, there's little you can do to make it proof against time. The best you can feasibly aspire to is to have the building be more of an asset than a headache. To that end, you probably want to make the girders pressure-treated too, although they needn't be rated for ground contact. The same holds true for the joists and boxing. Even though they're not in contact with the ground, the girders, joists, and boxing of a building set on posts are in contact with the elements, and to funguses or wood-munching insects they can look like so much chocolate cake. Still, to reverse myself, if you can get ordinary lumber much more cheaply than the pressure-treated kind and money is a problem, go with ordinary lumber and don't worry unduly. You can always paint your lumber with a clear wood preservative, which will protect it some, or you can let the decades roll by, watch the ordinary lumber turn a pleasant shade of silver-gray, and very possibly luck out. If that's what you opt for, add a roll of 8″ aluminum flashing to your order (flashing is simply light-gauge sheet metal packaged as a rolled-up strip), because we'll touch on making little termite shields for the tops of the posts. You'll need flashing later anyway, and while debate continues to rage (all right, maybe not in *your* circles) as to whether termite shields do any good, making them may salve your conscience, whose soundness is vastly more important than that of the outbuilding.

3. Truss Plates

To hold the girders to the posts you'll need truss plates, which are galvanized-steel rectangles with predrilled holes to drive nails through. The most convenient size is 4x9, but if your girders are in one piece you can go a hair smaller without any catastrophic sacrifice of structural integrity. If your girders are in two pieces you might go an increment larger to 5x10, since the truss plates will have to serve the additional function of strapping the butted girder halves together. The reason truss plate size isn't *too* crucial is that gravity does a good part of the work here anyway. Get eighteen truss plates, two for each post.

Girder, post, and truss plate

4. Nails

The subject of nails often provokes extensive opining among carpenters, most of which has never struck me as very interesting. Nails come in a variety of shapes, but the only kind you'll need to frame your outbuilding are *common* nails. Common nails can be had uncoated, galvanized, or covered with various plastic resins. The last kind are sometimes called "sinkers" be-

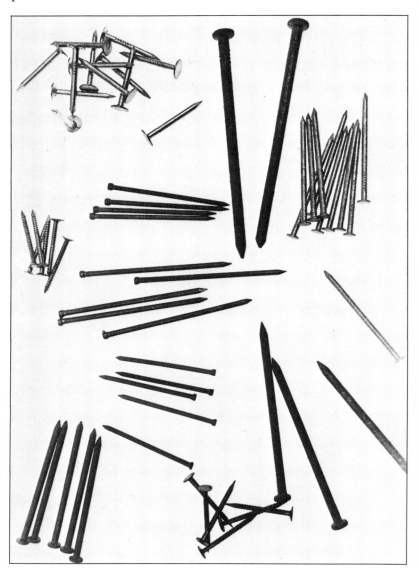

cause the plastic lubricates the passage of the nail into the wood. You can use them if you want; I have no stance on them one way or the other. Galvanized nails don't rust very readily, but the zinc coating forms a crackle finish that makes them slightly harder to drive than uncoated nails. You might want to use them where you're concerned about avoiding those little drip lines of corrosion under exposed nailheads, but for most purposes ordinary uncoated, unplated, smooth-shafted common nails will do perfectly well.

As with other types of nails, the size of common nails is specified in pennies, that archaic designation whose abbreviation is "d." The following table translates pennies into inches:

3d	$1\frac{1}{4}''$
4d	$1\frac{1}{2}''$
6d	$2''$
7d	$2\frac{1}{4}''$
10d	$3''$
16d	$3\frac{1}{2}''$
20d	$4''$

You might start by getting 5 pounds each of plain 10d and 16d common nails and a couple of pounds of galvanized 6d common nails. Sixpenny nails are a good size for nailing truss plates to wood, and since the bright surfaces of the truss plates can become miniature billboards for rust they're one place you probably want to avoid flaunting your corrosion.

5. Joists

As for joists, so long as your outbuilding is 16' wide or less you shouldn't have any problem finding 2x6s, pressure-treated or not, that span its entire width. Sixteen-, twelve-, and eight-footers are all widely available. You'll need three joists for every 4' of building length plus one extra joist to close up the last bay, so if your outbuilding is 16' long you'd order thirteen 2x6s, pressure-treated preferred, untreated Douglas fir next choice.

If, for whatever reason, you can't get 2x6s whose length equals the width of your building, figure on using twice as many joists, each of them half the width of the building. But also account for the need to tie the half-joists together with scabs.

Single scabs, each of them 2' long, ought to suffice for the purpose, so, instead of simply doubling the number of joists you need and ordering that many half-building-width 2x6s, add 2' to half of your order. That is, if you can't find, say, thirteen 16' 2x6s and have to make do with twenty-six 8' 2x6s, order thirteen eight-footers and thirteen ten-footers. That way you can cut a 2' scab off each of the ten-footers and wind up with the right number of joists and the right number of scabs. Moreover, since the scabs don't have to be *exactly* 2' long, you can save yourself some sawing by cutting the scabs so that the rest of the board is just the length you want. Presently we'll get to what that length should be.

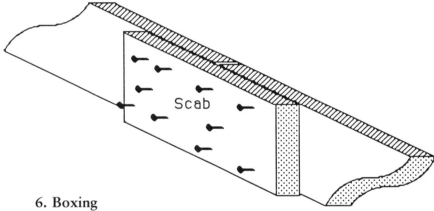

6. Boxing

For boxing, order 2x6 or 1x6 boards whose combined length equals twice the length of the outbuilding. The two joists at either end of the deck will essentially be their own boxing, so it's really just the joist ends along the two long sides of the building that you'll be boxing in. Obviously it's easiest to get a pair of boards each of whose length equals the length of the building. But if that's not practical for any reason, don't be especially concerned. If you go with pressure-treated wood—and I mildly suggest that you do—you may discover that you *have* to get 2x6s. Pressure-treated 1x6s are harder to come by. Generally 1x6 lumber is sold in the form of pine shelving, and if that's what you wind up making your boxing with, I'd recommend that you eventually paint it, since pine has a way of getting murky with mildew.

7. Styrofoam

Regardless of whether you decide to insulate the walls and roof of the outbuilding, I think you'll almost certainly want to insulate the floor (unless you plan to use the outbuilding exclusively as a storage shed). Floors can get cold even on chilly summer nights, especially when there's no basement or crawl space to protect their undersides from wind and dampness. The best material to use is Styrofoam, which, like plywood, is sold in 4x8 sheets. Unless you're building where winters are particularly cold, 1" Styrofoam should do the job, and if your winters do get bitter, go with 2". The difference in cost isn't that great. Either way, the thickness of the insulation isn't as crucial as it is for roofs. Heat escapes upward much more readily than it does downward. What we're mostly doing here is supplementing the insulating properties of the soles of your shoes.

Order enough 4x8 sheets to cover the entire floor of the outbuilding. Since you've designed the building in modules of 4', it shouldn't be very hard to calculate how much you'll need.

8. Plastic Vapor Barrier

To keep exterior moisture from wafting upward into the outbuilding you'll need some plastic vapor barrier. Hardware stores frequently have sales on 4-mil sheets 10' wide and 25' long. The kind of plastic isn't all that important. Some kinds deteriorate much more rapidly than others when exposed to light, but yours won't. Don't go with a thickness of less than 4 mils, though. Something thinner may be just as proof against water vapor, but it'll turn out to be unwieldy to put on. And make sure to get the kind that's transparent. Black plastic *works* just as well, but after you've tacked on the vapor barrier you'll want to be able to see the lumber you've tacked it to, or else nailing anything on *top* of the vapor barrier will involve rather more guesswork than is a good idea. You can also buy vapor barrier at the lumberyard, since nowadays it's routinely called for in new construction. As for quantity, buy enough to cover the deck allowing a few square feet for waste.

9. Plywood Decking

Use $\frac{1}{2}''$ or $\frac{5}{8}''$ plywood for the subfloor decking. If you plan to install carpeting or vinyl tiles, you'll have to put underlayment over the subfloor, and as underlayment chipboard or even particle board will do nicely. You can also have a perfectly presentable floor by laying plywood—the kind with one smooth side (or one "good" side, as they say in the trade)—and then painting it. For the subfloor, however, you don't need *any* good sides. Whatever knotholes there are will ultimately be covered by wood flooring or underlayment or more plywood. So just get plain old construction-grade exterior plywood decking. (Plywood is coded with letters that describe the condition of its surfaces: "A" means smooth and paintable, "B" means smooth with knotholes replaced by repair plugs, "C" means not so smooth with knotholes up to $1''$ in diameter, and "D" means not so smooth with knotholes up to $2\frac{1}{2}''$ across. Plywood that's smooth on one side but has scattered knotholes on the other would typically be labeled "AC." Rougher stuff might be labeled "CD.") As a general rule, there's nothing to be gained by using fancier materials than a particular task requires. It's better to save the money for when you *do* need fancier materials.

As for how much plywood to order, know that it comes in 4x8 sheets and order enough to cover the deck. The 4' modularity of your dimensions should make the computation a breeze. In fact, since you'll be using as many sheets of plywood as of Styrofoam, you've already made it.

10. Shims

You're going to be needing shims—thinnish sticks of wood to compensate for minor differences in the final height of the posts after you cut them. The most convenient shims you'll find are wood shingles, most commonly of cedar. Shingles are sliced on a taper; they're about $\frac{1}{2}''$ thick at one end and maybe $\frac{1}{8}''$ thick at the other. And since the taper is continuous, all thicknesses between $\frac{1}{2}''$ and $\frac{1}{8}''$ are available for shimming. You won't need perfect shingles to use as shims (a bundle of new ones costs around $20). So ask the people at your lumberyard if they have any beat-up rejects lying around (they often do), or scavenge a few old ones at your local dump. A dozen should be more than enough.

11. Scrap Lumber

And while we're on the subject of dumps, drop by yours to see what scrap lumber you can find. Longer boards are better than shorter, and appearance doesn't count. You won't need the boards for the deck, but you'll need them later for the walls and roof—not as structural members but as temporary braces and as poles to fish overhead members into place. The more scrap lumber you collect the better, so you might as well begin now.

Materials List

If all's gone well, you've just created a list of materials to take to the lumberyard. You place your order, and soon afterward you get your first big lumber delivery. The outbuilding has been mainly an idea thus far, but now that the delivery truck has come and gone and you're confronted with a humongous pile of wood, it's perfectly reasonable to allocate a moment or two for muttering, "Oh, my God! What have I gotten into?" Nevertheless, don't be overawed. The 6x6 girders and posts are the most massive materials you'll use throughout the project. The others will be matchsticks by comparison. And, as I've already said, massive as the 6x6s are, they're also about the easiest boards you'll work with in that they shrug off virtually all carpentry errors (except cutting them too short; that they don't like). So if you feel the need to be crude and carefree (within humane limits) about carpentering your newfound trove of lumber, now's the time.

BUILDING THE DECK

Once you've cut the posts you'll be off and running. Presumably you've retained the results of the post-height computations you made earlier. You added 6″ or so to the length of each post to obtain some working latitude, and now is when you use it. Now is also a good time to get hold of a carpenter's square—a graduated L-shaped piece of steel designed to provide you with all the right angles you'll ever need.

Rough-cutting the Posts

Mark off the posts on the 6x6s they'll be cut from, using the square to keep the lines perpendicular to the edges of the board. If you're cutting with a $7\frac{1}{4}″$ circular saw, you won't be able to make a deep enough cut to lop the posts in one stroke, but that's fine. Rotate the 6x6 and do it in two or three strokes. If you can't cut quite to the core of the 6x6 no matter how many times you rotate it, cut the cores with a hand saw. Or, easiest of all, borrow a chain saw. Whatever you do, make sure not to frolic in the sawdust. The stuff with which the lumber has been

All nine posts

pressure-treated is a compound of arsenic. Work outdoors, don't lick your hands, consider using gloves and plastic goggles, and urge your pets and children not to graze on the debris.

Cutting True Lengths

When you've rough-cut the posts, plunk them into their corresponding postholes, once again being careful not to let too much caved-in dirt sully the pristine tops of your newly hardened footings. This time, using the longest boards you have around instead of twine, begin marking what will be the true tops of the posts. Temporarily tack the boards to the posts with nails (when you "tack" with nails, you don't hammer them in all the way; you want to be able to get your nail-puller under them easily so you can pry them out), get your 4' level, and play around until you're absolutely sure that the path from the top mark on any post to the top mark on any of the other posts

Cutting off the top of a post

proceeds along a level line. Your lumber delivery will probably have provided you with boards long enough to let you level three post heights at a time. Don't use for leveling purposes boards that are so bent (start saying "crowned" instead of "bent") that they'll throw your level off. And don't take a whole day to do the leveling. If the top of one post winds up being $\frac{1}{2}''$ lower than the top of another post 16′ away, no posse of Error Troops will come to drag you away. You'll simply have to use a shim.

Now that you've marked the true tops of your posts and ascertained that all the marks are in a level plane, invent a labeling scheme that links each post with the hole it was marked in, label the rough-cut posts accordingly, and cut them to their true length. (Making the rough cuts has undoubtedly taught you everything you need to know about cutting posts.) Then put the posts back in their holes, but don't backfill. That comes later.

Squaring up the posts

Girders

While you've got the feel of cutting 6x6s, cut your girders to their exact length. Even if your outbuilding is 16' long and you've ordered 16' 6x6s, some of them may be a little longer. So get out your tape, mark them carefully, use your square to scribe the pencil lines, and cut the girders to their precise length. If you're making your girders in two equally long pieces, cut all the pieces now. And do it fairly carefully. If you're sloppy with a circular saw, direct your sloppiness toward cutting off too little rather than too much. As one carpenter put it, "Dead wood don't grow."

Leveling and Trueing the Girders

Next, lay out your girders in the position in which they'll finally be nailed, one girder atop each triad of posts. Using the dimensions in your drawing, mark the precise points on the girders where they intersect the post tops, then set them up so that the posts meet their marks on the girders. Now look along the girders for crowning; the chances are that none of them will be completely straight. Rotate them so that the crowns are in the horizontal plane—i.e., so that they bend sideways, not up and down—and you'll be less likely to wind up with unwanted hills or valleys in your floor. Be aware, though, that 6x6s can be crowned "both ways," sideways and up and down. If that's the case, rotate the girder so that (1) the bigger bend goes sideways, (2) the smaller bend goes up and down and (3) the center of the girder is slightly higher than its ends. When a girder is crowned from left to right, you can easily accommodate the bend by moving the middle post a little bit so that it falls under the center of the girder. But there's a limit to how much you can reasonably shim a girder if its center is very high. I'd say that if you have a girder that's crowned both ways and the peak of the lesser of the two crowns is more than 1" higher than its ends, exchange the board for a better one. Nonetheless, severe double crowning is comparatively rare and, under normal circumstances, the whole rotation procedure oughtn't to take more than a few minutes. If you're using half-length girders, butt the machine-cut ends together, prop them up so that the butt joint is just over the center of the middle supporting post, then strap them temporarily together by tacking short scraps of wood to their sides. (Once again, by "tacking" I mean nailing without driving the nails in all the way and using only as many nails as are absolutely necessary.)

Now, by sheerest trial and error, go around from post to post, level in hand and shingles at the ready, and do the final touch-up leveling of the girders. If any of your posts is a bit too low, split off a $5\frac{1}{2}''$ (approximately) piece of shingle and shim the post with it. If any of the posts is too high, remove it, trim off the extra $\frac{1}{2}''$ or so, put it back into its posthole, and try again. And while you're touching up their heights you should also be manipulating the posts to keep them as plumb (vertical) as possible. If you trust your ability to determine the vertical by eye, do that. Otherwise use the plumb bubble gauges on your level (the gauges that are perpendicular to the level). If the posts keep toppling over in their postholes, brace them temporarily with sticks. And remember that leveling each of the three girders isn't enough. They also have to be level with respect to one another. That means you have to look over your 2x6s until you find one that's not visibly crowned, lay it crosswise atop your girders (not on its flat side), put the level on the 2x6, and then do whatever shimming of the posts you have to do to get the 2x6 level. If you're using two-piece girders, there's an additional heuristic dimension: you'll want to stretch a length of twine along the entire girder to ensure that the two pieces form a straight line. And while you're doing all that, you have to see to it that the ends of the girders form a straight line (keep checking them with an uncrowned joist), that all three girders are spaced according to your deck diagram, and that the whole array forms a true rectangle as opposed to any mere rhombus. To ascertain rectangularity, use your long tape to measure both diagonals of what you hope is a rectangle, making sure you've used corresponding reference points in both cases. The best reference points are the outer corners of the outer girders. If you've got a rectangle, the two diagonals will be equal. If you don't, they won't. So play around until they are.

Two girders in place

Trapped forever in heuristic purgatory? Not at all. Setting up the girders won't take as long as all that. The posts don't have to be exactly centered on their concrete pads, and your shingle shims will give you ample room for fudging. If one end of the subfloor turns out to be $\frac{1}{2}''$ lower than the other, no one—probably not even you—will know. Discrepancies of $\frac{1}{2}''$ or less simply tend to vanish in a structure of this size. Yet don't be too careless, either. Scruples observed at the deck stage will redound to your benefit when you're framing the walls and roof. Deciding how perfectionistic you need to be comes down to balancing your intrinsic craftsmanship with your natural impatience. Both are real, and both deserve respect. So be punctilious but not lapidary. Trust your judgment.

Termite Shields

When the tops of your girders form a level, plane, rectangular surface, you're ready to yoke the girders to the posts with the truss plates. If you're using non-pressure-treated girders, however, there's a parenthetical procedure you'll want to attend to first: termite shields. So find the 8″ flashing you ordered, and with tin snips or even heavy scissors cut nine 8″ squares of aluminum and fish one in over the top of each post. Maneuver them until they're centered on the post tops and bend their edges downward along the sides of the posts. Done! And you've accomplished two things here. Not only have you thwarted termites (maybe), you've also flashed the girders, thereby preventing the posts from wicking rot-inducing moisture into the joint they make with the girders.

Yoking Girders to Posts

Now go ahead and lash the girders to the posts with truss plates, one on either side of each post. Set the plates up so that half is up against the post side and the other half is up against the girder, and use the 6d galvanized nails we talked about. Truss plates are generally drilled with a lot more nail holes than you'll need; nail them mostly around the edges. If your girders are in two butted halves, the truss plates on the three center posts will also hold both halves together. How they do that is self-evident. Just make sure to keep the girder halves butted tightly while you nail the plates.

Truss plate connects girder to post

Cutting the Joists

At this point you have three unconnected units, each of them a girder sitting on three posts. The joists will tie the units together, so begin by taking pencil and carpenter's square in hand and marking the joists where they'll have to be cut off. To wind up with a deck that's the size you want, the length of each joist has to equal the width of the building (some exact multiple of

4′) minus the combined thickness of the boxing. If you're using 2x6 boxing, subtract 3″ (the combined thickness of two 2x6s) from whatever multiple of 4′ you've chosen. If you're using 1x6 boxing, subtract $1\frac{1}{2}$″, since 1x6s are $\frac{3}{4}$″ thick. Once you've marked the joists, simply cut them to length.

Installing the Joists

In your sketch of the deck you've indicated the precise spacing of the joists along the girders, so carefully mark the points on the girders where the joists will be nailed. You can scribe two parallel lines, $1\frac{1}{2}$″ apart, where each joist will fall on each of the three girders it crosses, or you can take the traditional approach and draw a single line where one edge of the joist is supposed to fall and an X just beside the line such that the joist will cover the X when it's nailed. The systems work equally well, but since the second way is used more widely it would be the better choice if you expect to quit or abscond or die before the building's finished, in that a pro could more easily take things up where you left off.

After all the marks are on, use your 4′ level to check the joist spacing. In your diagram you spaced the joists so that if the outer 4′ edge of any sheet of plywood is flush with the edge of

Floor joists and boxing

the building, its inner 4′ edge will fall over the middle of a joist, enabling you to nail two butted plywood edges to that same joist. So if you line up one end of your level with the outer edge of either of the building's two end joists and set the level parallel to the girders, the other end of the level ought to fall more or less on the center of a joist. If it does, everything's gone according to plan. If it doesn't, go back to the chapter on drawing the drawing and correct the spacing of the joists.

Toenailing

Now find your 10d nails and toenail the joists to the girders. Start by nailing the two end joists; you'll nail the ones between after backfilling the postholes. If you're concerned that you haven't cut your joists so they're all of equal length, find two that *are* equal and put them on the ends, taking care that their midpoints fall precisely over the center of the middle girder. And if you're not comfortable with toenailing, now's the time to *get* comfortable. Consider this something of a practice run; once the outbuilding's built, no one will be able to see how egregiously you've screwed up or how much perfectly good lumber you've hammered into ugly, pathetic splinters. Try angling in each nail so that half of it catches the side of the joist and the other half sinks into the girder.

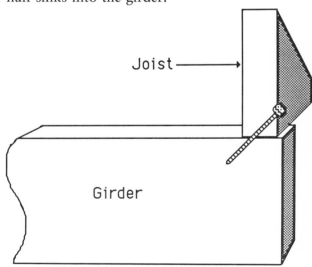

"Slice" showing joist toenailed to girder

While you're hammering the first nail, the joist will try to creep along the rafter, so, without putting a nail through your hand, hold it back. It will try to move anyway, but a certain amount of drift is okay, since after you put a nail in one side of the joist, you put the second nail in the opposite side, and nailing the second nail will tend to hammer the joist back to its marks. A third nail in the first side should do the job for each joist-to-girder joint. If after you've toenailed the third nail the joist still isn't on its mark, bang it with the hammer until it is.

Rechecking Squareness

Pause after you've nailed the two end joists. The deck still isn't very rigid, which makes this a good time to do one last test of rectangularity. Remeasure the diagonals of the deck and play around with it until they're equal. When you're satisfied that your deck is a level rectangle and that the posts are plumb and sitting comfortably on their footings, go ahead and backfill the postholes. That done, you can now nail the rest of the joists without being concerned that you'll throw the deck out of kilter. You may want to run a taut length of twine between the *ends* of the end joists to help you line the other joists up. That way at least one lengthwise edge of the building will be straight (provided you don't let any of the joist ends dent the twine). As for the other lengthwise edge, you can test for straightness either with twine or lengths of boxing, but wait until all the joists are nailed. If any one of them is egregiously overlong, trim it off with a saw. Too short is less of a problem. No one will ever notice.

Scabs

For those of you who are using half-length joists linked with scabs, the fastest way to get done is to nail the scabs on first, thereby manufacturing full-length joists. Use 10d nails (eight per scab should do, four into one half-joist, four into the other), taking care to keep the ends of the half-joists butted tightly together and the resulting full-length joist straight. Toward the latter end, a good straight reference board is helpful. And when you nail the put-together joists to the girders, make sure the butt joint of each one is just over the center of the middle girder. The scabs are at best a source of auxiliary support; you want the inner ends of the half-joists to be sitting on a couple of inches of good solid wood.

The Boxing

Now nail on the boxing, cutting it first if you have to. A pair of 10d nails through the boxing and into every joist end should do the job. Make sure the upper edges of the boxing are flush with the upper edges of the joists. You'll be nailing plywood to both, so you want a good flat surface to nail to.

That done, you should now have (at risk of belaboring the point) a sturdy level rectangle whose outer width and length are (1) multiples of 4' (within $\frac{1}{2}''$ or so) and (2) the dimensions you've chosen for the outbuilding. If that's what you've got, proceed. If it isn't (and often it isn't quite), fix it. As simple as that.

Installing Rigid Foam Insulation

After doing what you've just done—moving heavy boards and then smashing the living daylights out of them—installing the insulation, if you choose to do so, will feel almost like white-collar work. Start by scribing lines along the sides of the joists, keeping them parallel to the tops. Draw them 1" down from the tops if you're using 1" Styrofoam, 2" down if you're using 2". If you have scabs on your joists (no mean medical curiosity), draw the lines on the scabs too. Then, starting at the ends, drive 6d nails partway into the joists along the lines. Let the nails stick out about 1", and space them a "long" foot apart.

(A comment here about "long" and "short" measurements. First, the concept isn't one I just made up. It's widely used by carpenters, and it's valuable even when you're communicating only with yourself. The concept has two somewhat different senses, each one plain from its context. What's meant by putting in a nail every long foot is something like this: without using a ruler, guess at how long a foot is. Your mind will promptly furnish you with a multiplicity of guesses: short ones, long ones, and a slew of guesses in between. To get a long foot, simply

Cutting rigid foam insulation

select a guess that you're reasonably sure is on the long side without being *wildly* long. The concept's second sense usually applies to sawing boards. You take a measurement, and it's, say, $43\frac{1}{2}''$ *minus a few hairs*. You don't want to bother with measurements in 32nds of an inch, even 16ths of an inch. Probably your saw won't cut that precisely anyway. (We'll tactfully decline to explore the question of whether it's your saw or you personally that can't cut to close tolerances.) So you call the measurement a "short" $43\frac{1}{2}''$. What that means in practice is that you mark the board at $43\frac{1}{2}''$ and just barely saw off the pencil line. If it were a "long" $43\frac{1}{2}''$ you'd leave the pencil line unsawn. Of course, for an exact $43\frac{1}{2}''$ you'd saw off half the thickness of the pencil line and leave the other half on the edge of the board. (And, strangely enough, when you're about halfway done with the outbuilding you'll probably be able to do that.)

A reminder: if you've put a line of nails into the outer sides of either of the end joists, pull them out. Since their purpose is to hold up sheets of Styrofoam between the joists, the outer nails will tend to catch on your clothes and otherwise do little good.

Now pick a windless day (closed-cell Styrofoam is incredibly brittle; even a light breeze can tear it up) and cut the insulation into strips that just fit between the joists. A sharp kitchen knife is a good tool to use. Most of the strips will be $14\frac{1}{2}''$ wide, but where you've varied joist spacing as we've discussed they'll be a little narrower. If any of the pieces do break—and they will—

Fitting Styrofoam between the joists

just fit them back together and keep going. Unbroken pieces don't have that much more insulating value than ones with hairline cracks, and since the pieces fit snugly back together, hairline cracks are all you'll have. Set the strips down onto the nails, but don't push too hard. The final push will come from the plywood decking that follows.

Vapor Barrier

You're ready now to tack your sheet-plastic vapor barrier over the top of the deck. Where the edges of the sheets fall across the deck's surface, overlap them a few inches with the adjoining sheets. You can use staples or small nails, but staples are best. In fact, arranging that a staple gun be handy is a good idea. You'll want it later anyway. Whichever you use, don't put in too many. The plywood decking will ultimately hold the vapor barrier in place; the staples are mainly to keep the wind from taking it to China before you nail on the subfloor. A few scrap boards laid atop the deck should anchor it just fine. And don't worry now about any vapor barrier that droops over the sides. It's easiest to trim after the plywood is on.

Subfloor

Which brings us to the subfloor. If you've been assiduous about your deck drawing, you'll already have sketched the "brickwork" pattern of plywood sheets. The objective is to use as many uncut 4x8 sheets as possible and still preserve the alternating scheme. For the 16' by 12' outbuilding shown herein, only one sheet needs to be cut in half. Start by laying out all the plywood on the deck before you drive any nails. Where there are errors in length or squareness, array the plywood so that you split the resulting differences between two sides. Butt machine-cut edges to machine-cut edges wherever possible and, using your 6d nails, nail the plywood down to all the joists, putting the nails in at intervals of about a foot. If you're bad at estimating the paths of the joists under the plywood, keep doing it until you get better. A few nails hammered into air (actually into Styrofoam) won't matter one way or the other. And when you cut plywood, regardless of whether you use an all-purpose cutoff blade or a plywood blade in your circular saw, adjust the depth of the cut so it's fairly shallow. You'll use a full-depth cut for sawing structural members, but plywood does considerably less splintering if you take the time to find the appropriate wing nut on your saw and shorten the cut. You'll also find that when you need to draw lines on plywood your 4' level makes a good straightedge.

As soon as the plywood is nailed down, you'll no doubt want to walk around the deck and congratulate yourself. You're now

The deck

a carpenter—of sorts. You probably even have some inkling as to whether you're a good one. If you are, that's all to the good. If you're not, in a curious way that's even better. Except for a few details like walls and roof, you've built a building, and you've demonstrated that it doesn't take all that much skill to do it. Besides, incipient skills are bubbling up in you, and soon enough you'll have the chance to use them.

5

About Windows and Doors

Even though your building juggernaut is by now hurtling headlong toward completion, I ask that you tolerate an interruption. It's the walls that go up next, and before you build them you have to plan for what goes into them.

Whatever you imagine windows or doors to be, it will serve the matters directly at hand if you conceive of them as rigid, flattish, rectangular, wood-encased boxes that fit into correspondingly rectangular holes in a wall frame called "rough openings." What goes on inside the box is up to whomever made the door or window. *Your* job is to provide the appropriate rough opening, to line it with plastic or aluminum flashing (joints around windows and doors are moisture traps), to nail each box securely without smashing it to pieces, to fit siding neatly up against the outside edges of the box and caulk the seams, and, if you're finishing the inside of the building, to bring your Sheetrock or wallboard up to the *inside* edges of the box and trim the seams with some kind of molding. The briefest glance at a properly installed door or window in a frame house will give you an idea of what's required.

Get a feeling for the doors and windows available to you by obtaining some catalogues from your lumberyard or from one of the window outlet stores that seem to be proliferating in the suburbs and exurbs. (Window outlet stores will often give you better deals than lumberyards; since windows and doors are all they sell, they're apt to be less complacent.) Prices vary enormously from brand to brand, but I'm guessing that for an outbuilding you'll be buying mostly low-end stuff, so you may as well stick to low-end catalogues. Interestingly, low-end windows don't look as different from high-end windows as their prices might suggest. If you compare new wooden windows at opposite ends of the cost spectrum, the sashes (i.e., the frames that the glass is set in) will probably be made of the same clear ponderosa pine, and the glass will be equally transparent. Most of the difference will be in the way the units operate and seal. Expensive double-hung windows (the kind with upper and lower sashes that slide up and down in their tracks) will have more elaborately manufactured jambs and counterweights; they slide more smoothly and permit less air infiltration. The same holds for expensive casement windows—they crank more easily and seal more tightly—and, in the way they operate and seal, also for doors.

Nevertheless, in an outbuilding you may not care that much about smoothness of operation. Cheap windows may not open with the viscous uniformity of resistance you get from expensive windows, but they open. As for tightness of seal, smallish one-room buildings are amazingly easy to heat and, if anything, suffer chronically from closeness, so a little air infiltration is often a blessing. It's not that I mean to advocate junk. If Volvo begins making windows and you're wedded to Volvo, then by all means go out and buy Volvos. All I'm saying is that dollar for dollar, high-end windows won't give you nearly the same payoff in your outbuilding as they would in your house. And keep in mind that one large window will cost considerably less—and be much less trouble to frame out and install—than two small ones.

Returning to catalogues, you'll find that the unit numbers most companies assign their doors and windows also specify their size. I mention this because the practice is so widespread that the catalogue may not tell you it's so. Generally the unit number gives the width and then the height of the window opening if you were to install the window and then remove the sashes. But there's another squiggle. A typical unit number might be 2846, but if you rush to conclude that the window in question is 28″ wide and 46″ high you'll set the guy at the lumberyard clucking about the "feller come in here, got all them fancy degrees, and he's complainin' about how some 2846 winder he bought ain't 28″ wide and 46″ high." To which his cohort may ruefully reply, "Validates everythin' I been sayin' about them glarin' lacunae in today's graduate education." What's meant by 2846, you see, is 2′8″ wide by 4′6″ high. And generally you say not "twenty-eight forty-six" but "two eight, four six." The standard height of most doors sold for new construction nowadays is 80″ (i.e., 6′8″). They're consequently called "six eights," meaning that anyone up to 6′8″ tall can walk through them without skimming the top of his head off.

Catalogues also specify the rough openings required for each model of door and window and, for windows, also indicate if they're available single-glazed, double-glazed, triple-glazed (those refer to layers of glass with air seals in between), or insulation-glazed, the last made with special types of glass that purportedly control the flow of radiant heat so you're not too hot in summer or too cold in winter. It's my own feeling that anything fancier than double-glazed is extravagant for an outbuilding, and unless you're especially concerned about winter condensation on the insides of the windows (rarely a problem in a structure without moisture-generating kitchens and bathrooms), the less expensive single-glazed windows ought to do perfectly well. Again, it's not just out of fondness for the inexpensive that I say this. No matter how well-fenestrated they are, small, one-room buildings can get sepulchral, so you'll probably be keeping windows open a little anyway.

As for doors, the only suggestion I have is that you use an exterior-grade entry door so the weather doesn't take it apart. Exterior doors come in steel, glass, wood, and combinations thereof, and new ones are usually insulated. If you like one that isn't, I've already said my piece about overdoing weathertightness. Use it and don't worry about it. Admittedly, the door in the outbuilding we'll build in the following pages is a touch luxe. It's got two matched double-glazed glass units set in pine; one unit swings open, the other's fixed. Yours needn't be that elaborate. And if you do buy a new door, by all means get one that's prehung in its jambs (the jambs are the members that immediately surround the door).

If, however, you prefer to use an old door, you'll have to hang it in jambs that you make yourself. You'll also have to know how big a rough opening to leave in your wall frame. The second part is easy. Convention correctly dictates that your rough opening be $2\frac{1}{2}''$ larger in both width and height than the door you're going to hang. That way there's room for the jambs (normally made of $\frac{3}{4}''$ pine), the door, and whatever shims you use. The first part—the actual hanging of the door—merits more explanation than we have room for here. Any number of home-improvement books will guide you patiently through the process, and if you have an old door you're in love with it's definitely a project worth taking on. Hanging doors isn't the baffling, arcane skill it's sometimes made out to be, and there's no particular hurry about it, either. Homemade jambs are seldom assembled around a door in freestanding fashion. They're generally built inside the rough opening, so you'll have to frame out your walls first anyway.

If it's only new windows you'll be working with, you can stop searching when you've found the right catalogue-cum-price-list. Otherwise make your pilgrimage to salvage stores, junkyards, dumps, whatever. But plan on installing at least one reasonably new window so you can have it around to see how the jambs and casings go together. Then you can restore your old windows so the woodwork more or less matches the new one. You may also have to repair some broken panes, but that's not especially difficult, and a good home-repair manual will give you all the guidance you need.

Before actually buying your door and windows, though, walk the deck you've just built to determine where you want the light to come from, where you *don't* want it to come from, and what it is you want to see through your windows. I needn't remind you that you'll probably want to install more fenestration in, say, a painting studio than a darkroom, but do be careful not to put in windows where what you really want is floor-to-ceiling bookshelves. An old rule that still has merit is to keep the windows in the north wall of a building small and few, since the most cold and the least light come from the north. Take the time to look at other windows and doors in existing structures, then rev up your imagination, plant yourself on your deck, and try them out in your outbuilding-to-be. If you'll be using a wood stove or in-the-wall gas heater, make sure to leave appropriate wall space. Have pencil and paper with you and sketch out actual dimensions. Now that the catalogues and salvage stores have made you something of a fenestration-ingress-egress savant, doors and windows are no longer just amorphous holes in walls. They've got dimensions and price tags too, the latter not only in money but maybe also in restoration time. So you've got some designing, compromising, and all-around parameter juggling to do, and this is the time to do it. It's also the time to fix any old windows that need fixing, because once the jambs are ready you can determine the rough openings they'll need. Simply measure the restored unit, add about $1\frac{1}{2}''$ to both height and width, and call that the size of the RO (an abbreviation you may see in catalogues).

In the end, you should have four sketches, one for each of the four walls of the outbuilding. They don't have to look pretty, since they're really just notations to yourself. The height of each wall (seen from the inside) should be scaled at 8′, the length of the two "long" walls should equal the length of the outbuilding minus 8″ (2 times $3\frac{1}{2}''$ to account for where the 2x4s of the adjoining walls butt in, plus another 1″ for wallboard or Sheetrock, paint, wallpaper, and error allowance), and the length of the two "short" walls should equal the *width* of the outbuilding minus 8″. The rough openings for the windows should be set at least 4″ in from the ends of the walls (to allow for window trim) and a short 15″ down from the tops (I'll explain why in the next chapter). That means the actual window openings will start about 16″ down from the top of the wall, so, knowing the window size (caution: the specified height, not the RO), you can estimate within 1″ or so how high off the floor the bottom of each window will be. Crank up your imagination yet again and this time furnish—yes, that's what I said—furnish the outbuilding in your mind. Remembering that desks and tables are normally about 29″ high, see if any of the window bottoms you've chosen are lower than the tops of nearby desks, tables, sofas, etc. Is this what you want? No? Then pick correspondingly shorter windows.

One final note. You may be protesting by now that the "short" walls of the outbuilding aren't the rectangles I've implied they are. True enough. They're actually pentagons—rectangles with isosceles triangles perched on top. And if you want lots of light, you can put small fixed-glass windows up in the triangles (they're called "gable ends"). Windows of that sort are available in a variety of shapes: circles, octagons, triangles, squares, what-have-you. We haven't yet determined the dimensions of the gable ends, but if you're bent on having light up there, read ahead to the chapter on framing the roof and pick the roof pitch you'd like. Once you know how your roof will be pitched, it's a fairly trivial matter to calculate the length of the legs of each gable-end triangle, at which point you can insert (on paper for now) whatever windows you like. Just make sure to leave a little room in the gable end for framing members. The bottom of the rough opening for these windows can be the same as the top of the double wall plate (a quick read-through of the next chapter will clarify all this), but don't plan on bringing the bottoms of these windows any lower than 8′ off the floor. That would mean cutting into the wall plates, which would make your carpentry complex to the point of impracticality.

Now, if you like, go ahead and order your door and windows. In the next chapter you'll build the walls that contain them.

6

Framing the Walls

THE LUMBER

What and How Much to Order

You face an option here. Choice one is to make scrupulously detailed drawings of the walls, then use them to compute how much lumber you'll need (only, in all likelihood, to be wrong anyway). Choice two is to invoke a rule of thumb, in which case if you're off by a lot, you can blame either the rule or, if you absolutely need to personalize, me. My own vote goes to the thumb. That way, if you end up with too many 2x4s you can always use them to frame the gable ends, while if you have too few you can go to the lumberyard, carry home whatever additional wood you need, and still be building where you'd otherwise be mired in sketches. Having the lumber at hand also permits the luxury of making life-size drawings with the wood itself. Should you need to visualize a wall frame, you can simply array the pieces on the deck and play with them until they look like the wall you want to build.

The only kinds of boards you'll use are 2x12s for headers (the horizontal members over the doors and windows) and 2x4s for everything else (studs, sills, plates, etc.).

To employ the rule of thumb for studs, start by doubling the width and the length of the outbuilding and adding the two figures together. The result is the number of linear feet of wall you'll be putting up, which should be an integral multiple of 4. For every 4′ of wall, order three studs. That's your starting figure, but it won't be enough.

So add four more studs, one to terminate each of the four walls with.

Then add another four, since you'll end two of the four walls not with single studs but corner posts, which you'll make by sandwiching triads of "cripples" (short lengths of 2x4) between two whole studs.

Then add still another four to frame out the rough opening of your door with.

Finally, add three for every window. Framing window openings (door openings as well) uses extra lumber, but it also saves lumber, since there's no framing inside the hole. So this last figure's really just a guess.

Write down the resulting number and order that many "precuts," those being 2x4s already cut to a length of 92 5/8″. When set up on 2x4 sills and topped off with 2x4 double plates, precuts give you ceilings that come in a shade over 8′ high. That's if you're making ordinary horizontal ceilings. The outbuilding we're building here has *sloping* ceilings, but so what?

Calculating how much 2x4 you'll need for sills and plates is easier. Your walls will have single sills and double plates, and you already know how many linear feet of wall you're building. Triple that figure, and you have your answer. Needless to say, it's most convenient if you buy the wood in lengths that come out evenly. For a 12′ by 16′ outbuilding, six twelve-footers and six sixteen-footers are just the ticket. But half-lengths are fine too; there's no special virtue in making your sills and plates out of whole pieces of lumber.

Headers, however, *do* have to be cut from whole pieces. So start by making a list of widths of each of your rough openings, including the door (but excluding any little windows in the gable ends; 2x4s are header enough for those). Add 4″ to each of the widths, since headers are a little longer than the openings are wide. Since headers are doubled, you'll need two lengths of 2x12 for each of those "width plus 4" dimensions. Now, given that

2x12s are sold in the usual lengths (8′, 10′, 12′, etc.), sit down and figure out how many 2x12s you'll need, and in what lengths you'll need them, to give you that many whole pieces with a minimum of waste. (Don't make them too long, though; 2x12s can be stunningly heavy.) Those plus the 2x4s constitute all the lumber you'll need to frame the walls.

All the lumber, that is, if you've succeeded in finding some long scrap boards at the dump. If you haven't, throw in a couple of extra sixteen-footers, and maybe a couple of twelve-footers too. You'll want them around to use as braces, struts, what-have-you, and you can be fairly sure that the outbuilding (autonomous organism that it sometimes seems to be) will eventually find a way to incorporate them into its innards.

What Kind of Wood?

One last note: As for the kind of lumber to use, my preference is for Douglas fir, which builders refer to as "Doug fir" or just "fir" or sometimes "fuh." It's pretty much become the nationwide standard for residential construction (although spruce, hemlock, and southern pine are often available too), and what makes it so appealing is that it's at once hard and very nailable. (Oak, by contrast, is *very* hard but not really nailable; when you drive nails into oak, they either bend and don't go in, or they *do* go in but split the oak en route, which accounts for why oak is generally drilled and then either screwed or pegged or otherwise joined in some nail-free way and also for why the use of oak in construction is largely limited to making posts and beams.) Still, despite my feelings about fir, spruce, pine, and hemlock will work too. We're really just talking about nuances, so, whichever you choose, you can be at peace with yourself over the choice.

THE WORK

Laying Out the Walls

Laying out the walls is the last serious design chore, and, as I've already noted, you can do it by putting pencil to paper, or you can let the deck itself become your sketchpad and move the wood around until it takes the desired form. You can also evolve some custom combination of both those techniques (perhaps sketching what's not so obvious and not bothering to sketch what is), or you can decide that your imagination has arrived at that pinnacle of fitness in which all you need to do is picture the walls in your head, write a few dimensions down, and go from there.

It's true that you could take a vastly different tack and diagram everything in triplicate, but that would fly flagrantly in the face of our initial approach. What we're after is a satisfactory outbuilding, not a journey through every last nicety of technique. And I'm aware that taking so expedient a position might seem to controvert a few millennia of oriental wisdom, in which, at least for the novice, the end result would be asked to play a muted second fiddle to the means. Yet for all its stubborn attachment to expediency, the West has managed to provide more and better shelter for more people than perhaps any other civilization in history, and since it's shelter we're after it would be imprudent not to tap the body of wisdom with the best relevant track record. If you come out of this experience still not quite able to meet every touchy situation with a deft flick of your samurai hammer, you'll at least have an outbuilding to take refuge in.

Stacking double plates and sill for stud wall

Cutting the Wood

So pick a wall to start with, and refer to whatever notes you've made on the rough openings that go into it. To make it easier to raise the assembled wall and then lug it into place and brace it once it's up, I suggest you do a long wall first. (The short wall in the photograph on page 93 is the exception that proved the rule.) The long walls should be 7″ shorter than the full length of the building (thereby leaving $3\frac{1}{2}″$ on either side for the thickness of the short walls). After determining what that dimension is, you can start by cutting and laying out two equal lengths of 2x4, one for the soleplate at the base of the wall and one for the uppermost of the two plates at the top of the wall. The lowermost of the doubled top plate is cut to the full length of the building, so it can lap over the adjacent short walls and "strap" them in tightly. If you're making the lowermost of the top plates out of more than one piece of wood, try to arrange that the butt joint between the two lengths falls over a door or window header. It's not essential, but that way you can nail down each of the butted ends to a header, and the joint will be substantially stronger. At the very least, the butt joint should fall over the middle of a stud, so there's something for the inner end of each piece of wood to be held up by.

All the plates and sills stacked in position

Marking the Placement of Studs

Now place the sill (or sole plate if you prefer) on the deck where you want the wall to go. Make sure it's centered on the deck so there's $3\frac{1}{2}''$ between each end of the sill and the edge of the deck. Once it's properly positioned you may want to put a nail through it to hold it temporarily in place. Next, mark the places on the sill where the wall studs will be nailed, making absolutely sure that all the studs stack directly over the floor joists. Actually, stacking won't be possible with the two end studs, since the joists at either end of the building fall beyond the edges of the wall. But it's possible—and essential—with all the others. Then center the lowermost top plate on the soleplate you just marked (thus leaving $3\frac{1}{2}''$ of overlap on each side), and transfer the marks on the soleplate to the top plate. Put them on as many faces of the wood as you have patience for. Too many marks, if they're accurate, are always better than too few. (If they're *not* accurate, you get this wonderful morass of confusion, tilted studs, and occasions to use your trusty nail-puller.)

Where you're putting up unfenestrated wall, you'll simply lay out precut studs between the sill marks and the plate marks, and, if you're building the wall on the ground, end-nail them with a pair of 16d nails. But nailing's for later. What you need to do now is get your ROs (rough openings) into the wall. So let's quickly go through the rudiments of window framing.

Calling the Members by Their Names

When you set a window into its rough opening, the horizontal piece of 2x4 it rests on is called, once again, a sill. The sill is 3″ longer than the width of the rough opening, and the cripple 2x4s that hold it up are called "jacks." There's one jack at either end of the sill, and there's a jack for every point between at which, if you weren't installing a window at all, there'd be a full-length stud. Rising from either end of the sill are 2x4 trimmers. Their length is exactly equal to the height of the RO, and the face-to-face distance between them is exactly equal to the *width* of the RO. Spanning the tops of the trimmers is the header, which is also 3″ longer than the width of the RO and which you'll make by nailing two equal lengths of 2x12 together.

Doors are framed very much like windows except that the ROs don't have sills or sill jacks. The trimmers come all the way down from the headers to the soleplate, which, to accommodate the door, is cut away so that the subfloor itself constitutes the bottom of the rough opening.

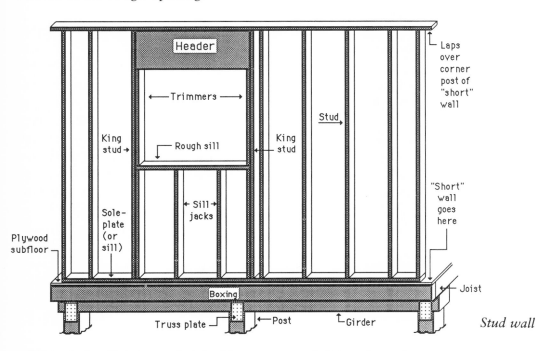

Stud wall

The whole affair is sandwiched between two full-length studs called "king studs." If you locate a window so that neither of its king studs stacks directly over a joist, you have to put in two extra precuts for the rough opening. But if you place the window so that at least one king stud *does* stack just over a joist, you don't have to double it. The same precut can be both ordinary stud and king stud. What this suggests is that you might want to move some of your windows a few inches to the left or right of where you originally planned to put them. Making studs serve double duty is a way of saving lumber.

Building the Wall Western-Style

There's nothing to stop you now from assembling your wall and nailing it together. In fact, if you've prepared and marked your plates and have your precuts around, the unfenestrated part of the wall is already cut. All that remains is to decide whether to build the wall flat on the deck and then hoist it up or to build it in place. You ought to be able to hoist up a 16′ wall with the help of one strong, competent, sensitive accomplice, and a wall as long as 24′ with two. Novices generally find it easier to build walls on the ground, but if you don't happen to know anyone who's strong, sensitive, and competent and don't expect to meet such a person by the time the wall is nailed together, build the wall in place. We'll go through framing the wall Western-style (that is, on the ground) first, so that we'll have, if you will, a frame of reference in which to discuss building the same wall in place.

Start by clearing the extraneous debris off the deck and arraying the soleplate and lowermost top plate so that they're (1) on edge, (2) more or less parallel, and (3) about 8′ apart. If either of them is composed of two lengths of 2x4, that's fine. They'll come together soon enough. You've planned your rough openings by now, so presumably you know, and have marked, where all the full-length precut studs go. So line them up between the soleplate and top plate, make sure they're on their marks, and end-nail them with 16d nails, two through the soleplate and two through the top plate. Try keeping the wall as rectangular as you can get it right from the outset. Yes, you can always square it up later, but there's no reason not to get off to a clean start. Also, figure out which studs will fall 4′, 8′, 12′, etc. from the corners of the outbuilding and make a special effort to ensure that they're as straight as possible. It's to the outsides of those studs that you'll be nailing the edges of your siding, and if they belly away from the true vertical you may find when you're nailing siding that, somewhere around the middle of the stud, you can't find enough purchase for both sheets. Attending to this now, when it's easy (albeit annoying), may spare you a lot of grumbling later on.

Headers for Doors and Windows

Well, if there are no doors or windows in the wall, you're done, since all its studs are full-length precuts. If not, this is a

Window header

good time to mark *exactly* where in the wall your door and/or windows go. You can see the wall *in toto* now and therefore make any last-minute adjustments in window placement. When you're satisfied, go ahead and cut your headers: two whole pieces of 2x12 per rough opening, each piece 3″ longer than the width of the RO. Use your carpenter's square carefully when you mark the cuts, then use your saw carefully when you make the cuts. You want them to be as square as you can get them. Then, if they're window headers, face-nail the pairs of 2x12s together with 10d nails. A nail near each corner and another nail for every foot or so of header should keep them adequately tight. And line up the edges before you nail the boards; each resulting doubled header should feel like a single piece of wood.

Using 2x12 headers gives you a "normal, natural" window elevation. But heading a door RO with 2x12s won't guarantee that you'll come out with a rough opening of the proper height. You may, but rough openings for doors are pretty variable, so before laminating the two header boards together it's possible you'll have to play with them a bit. I'm assuming that by now you know what size rough opening your door requires. What you have to fabricate is a header whose $11\frac{1}{2}″$ dimension is sufficiently augmented or reduced so that your door RO comes out to the proper height. Therefore, take one of the pieces of 2x12 that you've cut for the door RO and butt it temporarily into place up under the lowermost top plate. Then measure carefully from the bottom of the header board to the bottom of the soleplate. (Yes, the bottom; you'll eventually cut away the section of soleplate that crosses along the bottom of the RO.) If the resulting dimension is less than the height of your door RO, you'll have to "rip down" your 2x12s to get the measurement you want. Use your circular saw for this purpose, and don't be excessively punctilious about the cuts. Eventually the doorframe itself will hide the edges. If the resulting dimension is *greater* than the height of your door RO (which is likely to be the case if you're putting in a glass slider or patio door), you'll have to build up the header. You can do that with sections of 2x4 (on the flat), 1x4, or, for finer gradations, thicknesses of plywood. Just make sure that each "build-up" board is cut to the same length as the header itself and is $3\frac{1}{2}''$ wide. Now you can nail the header boards together, then nail the "build-up" boards to one of the long edges of the header.

Nailing the King Studs

Next, onto each fabricated header face-nail any king studs that aren't already joined to the wall frame. (We said that ordinary studs could also serve as king studs; any such double-

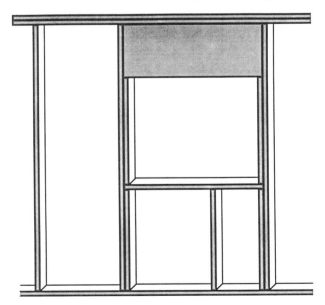

Window rough opening

duty studs will have already been nailed into the wall, so what we're talking about here are "pure" king studs—"pure" in that they don't fall on any of your 16″ centers.) Make sure the tops of the king studs are flush with the tops of the headers, and use six nails (either 10d or 16d) per joint, three into the end of each of the two lengths of 2x12 that make up the header. Nailing the king studs now gives you considerably more room to swing a hammer than you'd have if the headers were already nailed in place.

Terminology

(Good news and bad news about the growing profusion of terminology. First the bad news: it gets worse. The good news is that reading is a vastly different experience from building. When all the stuff is right in front of you, your inner voice isn't going to want to refer to a header as "that thick piece of doubled wood that goes on top of the window hole," much less to a king stud as "that long piece of 2x4 that comes down from the top of the wall, helps hold up the aforementioned thick piece of doubled wood from the side, keeps going down to the bottom of the window hole, helps hold up the piece of wood across the *bottom* of the window hole from the side, and finally goes all the way down to the bottom of the wall, where it rests on that long horizontal 2x4 that's nailed to the floor." I have more respect for your inner voice than that. Your hands and eyes will teach it all the terminology in this book plus some you improvise yourself.)

Installing the Headers

Now fit the header(s)-cum-king studs into the wall, jam any bare header ends up against the studs they'll be nailed to, and end-nail the king studs just as you did the ordinary studs. Finish the job by nailing any open header ends to the double-duty studs that hold them up, then face-nailing the lower top plate down onto the tops of the headers (one 16d per foot is okay, but if any of the headers has a top-plate butt joint over it, add an extra pair of 16d nails to secure the ends of the 2x4s).

If there are any butt joints in the soleplate, find a scrap of 2x4 to scab them temporarily together, and don't let the scabs come down below the bottom of the soleplate. You're doing this so the butt ends don't flap around while the finished wall is being moved.

Raising the Wall

The wall frame now contains all the full-length precuts it's going to get, and it's ready to be raised up. Its rough openings still need to be framed out, and the upper section of its double plate still has to be nailed on, but it makes sense to postpone those jobs until the wall is standing. Light walls rise as if borne up on zephyrs; heavy ones give rise to hernias.

Before raising the wall, though, prepare a couple of braces to hold it up after it's standing. Precuts will do; scrap lumber, if you have some pieces that are precut length or longer, is even better. Collect the friend or friends we've spoken of—the ones so selflessly preoccupied with *your* concerns that they can barely get on with their own lives—and slide the wall so that its soleplate is near the edge of the subfloor it will be nailed to. Stand up the wall, shimmy it into place (not so nightmarish a job as might seem to be the case in print), and fuss with it until the outside edge of the soleplate is decently flush with the edge of the deck and the ends of the soleplate are $3\frac{1}{2}''$ from the corners of the deck. As soon as you think you've got it, put a nail—*any* nail—down through the soleplate into the deck. Then do it with a second nail. You're at a milestone here, so you want to freeze it in place. The serious nailing can come in a couple of minutes.

When the wall is standing, it shouldn't take more than one person to hold it up while the second person hammers nails. So first nail a brace to each of the wall's two end studs. Use only a single 16d (allowing the brace to rotate on the nail), put it at least two thirds of the way up from the bottom of the stud, and nail the brace to the stud's inside surface (so it won't get in the way of the adjoining walls). Then nail a couple of short blocks of wood to the deck, one near where each of the two braces touches the deck when the wall is more or less plumb. That done, nail the lower ends of the braces to the blocks. At this point you and your affiliates ought to be able to walk away from the wall without its falling down. Inspect the ends of the braces for joints that seem to cry out for an extra nail and go ahead and give them what they crave. You may even want to add another brace or two if the wall seems to need it. Just try not to crowd your workroom on the deck. And don't overnail the braces or splinter any perfectly good 2x4s.

Now, once you've taken a moment or two to gaze with guarded satisfaction on what you've accomplished, nail the wall securely to the deck by putting a 16d nail down through the soleplate into each of the joists. You may complain here that all the joists have studs stacked over them, precisely as they're supposed to. Fine. Drive the nails into the soleplate just *next* to the studs, and pick an angle that's likely to sink their points into the joists.

"Short wall" in place; top plate not yet installed

Framing the ROs

Then, if there are any rough openings in the wall, finish framing them. If there's a door RO, cut a pair of 2x4s for trimmers (their length is the distance between the bottom of the header and the top of the soleplate), jam them into place, face-nail them to the adjacent king studs (one 10d nail every 16"), toenail a couple of 16d nails through the upper ends of the trimmers and into the headers, then toenail another couple of 16d nails through their lower ends into the soleplate. Now cut away the section of soleplate at the bottom of the RO. (Waiting until now to cut it has made it easier to hold the wall together while moving it around.)

Door rough opening

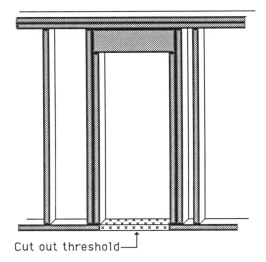

Cut out threshold

When you try making the cut with your circular saw, you'll soon enough realize that you can't do it. There isn't enough room for the body of the saw. So, just this once, use a handsaw. And I hope it's occurred to you not to nail down that particular section of solepate, because if it hasn't, you'll have to take the nails out. You can use a nail-puller for this, or once you've sawed through both sides, you can simply bash the bejeezus out of it with a hammer and blow away the splinters.

For window ROs, start by cutting the sill (width of RO plus 3") and the jacks, which run between the bottom of the sill and the top of the soleplate. You'll need one jack for every joist under the window (you stack them over joists just as you'd stack full-length studs) plus two more for the ends of the sill (unless, by sheerest chance, one of the end jacks falls just over a joist). Cut the jacks a little more carefully than usual. It's important that they all be the same length. Then take a pair of them, butt their bottoms tight against the soleplate, and face-nail them to the insides of the king studs (a 10d nail every foot). Now, after marking the RO sill so its stud marks correspond to those on the soleplate (you want your window jacks to be decently vertical), set it across the tops of the jacks and put two nails through the sill and into the end of each jack. Jam the rest of the jacks into place between the sill and soleplate, end-nail them from the top (two nails each), and toenail them at the bottom. If you've measured correctly, the distance from the bottom of the header to the top of the sill should equal the height of your RO. If it's off by a lot, correct accordingly. If it's off by 1″ or less, don't do anything until you actually try putting in the window. Rough openings aren't called "rough" for nothing. Permissible latitudes are fairly large.

Now cut the trimmers, wedge them up against the king studs between the header and sill, and face-nail them into the king studs (again, one 10d nail every 16″ plus a couple of toenailed 10d nails at the top and bottom). The ends of the header are now borne by the trimmers, which are borne by the ends of the sill, which is borne by the jacks. If the width of the resulting RO is reasonably on target, proceed to the next RO. If there aren't any next ROs you're done.

In fact, you've just finished framing a wall. A long wall, no less. So next take it a little easy and build a short wall. When you've finished it and butted it against the long wall, the two walls will brace each other up, and you can knock away one of the temporary braces that's doing the job now.

Building the Short Walls

You build the short wall just as you did the long wall, with these exceptions:

1. Where the long wall was 7″ shorter than the length of the outbuilding, the length of the short wall exactly equals the *width* of the outbuilding.
2. Where the long wall had single precut studs at either end, the short wall gets corner posts at either end. We'll presently discuss how to make them. (They're easy.)

3. Where the lowermost of the long wall's top plates was 7″ longer than the wall itself so it could reach over and grab the tops of both adjacent walls, the lowermost of the short wall's top plates is 7″ *shorter* than the wall itself. That leaves $3\frac{1}{2}″$ at either end for the long wall's top plates to fit into.
4. Where the uppermost of the two top plates of the long wall was 7″ shorter than the lowermost, the uppermost top plate of the short wall is 7″ longer than the lowermost. As you'll soon see, doing it that way will permit the uppermost top plates of all the walls to mesh at their ends in a way that helps hold all the walls together.
5. Where the studs in the long wall had to stack directly over joists, the studs in the short wall are set up somewhat differently. The first stud mark you put on the soleplate is at $15\frac{1}{4}″$. That places the stud so that its *center* is exactly 16″ from the end of the wall. The second mark goes 16″ from the first mark, the third mark 16″ from the second mark, and so forth. Installing the ordinary studs in this fashion will ensure that when you're nailing sheets of 4x8 siding, the lapped edges of each pair of sheets will fall over the same stud, so you can nail them together easily.

    As for corner posts, every carpenter has his own recipe, but what the way I'll recommend has going for it is that it asks for little in the way of lumber or labor and still manages to get the job done. All you do is sandwich three scraps of 2x4, each about a foot long, between two precuts and face-nail them well with 16d nails. Put one scrap piece at each end and one in the middle. Keep in mind that when you're wiring the building for power you'll have already nailed on the siding. You're going to have to run cable along the insides of the walls, and probably you're going to have to turn corners with it. Now, cable being anything but limp, fishing it around a corner is hard enough. But running it through a corner made of solid wood borders on—forget "borders on"; *defines*—the impossible. And since outlets are normally placed a little over a foot off the floor, that's the height

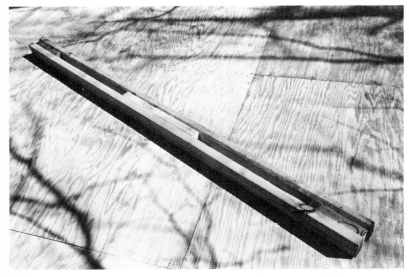

Corner post assembled

at which you'll be running wires. So when you drill into your corner posts to make holes for the wires, you'll be immensely relieved to find that the post is hollow inside and that you can shimmy the wire through.

Incidentally, you might be wondering why we use corner posts at all. I won't bother you with an extended explanation here, but, structural questions aside, think about the interior walls. Think about nailing wallboard or Sheetrock to the corners of the walls. It will soon enough become apparent that buildings without corner posts have incomplete interior corners, which makes it nigh impossible to nail the corner edges of the wallboard.

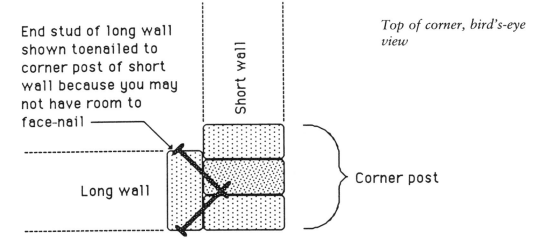

Top of corner, bird's-eye view

You know all you need to know now to build the remaining walls. After you've built the first short wall, stand it up, move it into place (no doubt your friends have been badgering you ever so gently to see when they could next be of help), get it under the top plate of the long wall, butt it tightly into place, and nail the top plate of the long wall down onto the short wall with three or so 16d nails. (This whole procedure will be an interesting test of everyone's mettle.) Then, because you're unlikely to have enough hammer-swinging room for face-nailing, toenail the end stud of the long wall to the corner post of the short wall (or, if you prefer, the other way around) using 16d nails every 16″. And don't forget to nail the soleplate to the joist running underneath. You don't want your nice new wall sliding around all over the place.

With a short wall done, I'd recommend going on to the next long wall. If you wait to build the second long wall last, you'll find there won't be enough room on the deck to make proper use of it as the mammoth worktable that it's been so far.

Assuming, that is, that you've been building Western-style and framing walls on the deck. If you've chosen to build Eastern-style—what the hell; let's call it that whether it's correct or not—and you're an attentive reader chances are you've already been transposing from the flat into the erect as we've gone along.

Eastern-Style Framing

Actually, the transposition isn't all that big a deal. You'd start by changing the order in which you erect the walls: doing both short walls first gives you the initial stability of corner posts and, when you need it, ready bracing for the long walls. For each short wall, then, begin by cutting the soleplate and lowermost top plate. Mark them for studs just as we spoke about above, then make the corner posts. End-nail the corner posts to the soleplate, tack on a couple of short diagonal braces to keep them square with the soleplate, stand the whole business up, and nail it to the edge of the deck where it belongs. Even if there's no one in the entire world who gives a flying fig for you or your projects, this is something you can comfortably do by yourself. Tack another couple of braces to the insides of the corner posts so they don't flop over, get up on a ladder, stretch the lower top plate across the corner posts, nail it down, and—lo!—you've outlined the wall in wood. (If the top plate is in two pieces, improvise a way to make it act like one piece; a scab will do nicely.)

Then install whatever full-length precuts go into the wall by end-nailing them at the top (two 16d nails) and toenailing them at the bottom (three 10d nails: two on one side, one on the other). When you frame out your ROs, you can still make your headers and nail them to their king studs on the ground, then simply pick up the whole affair, set it into place, and nail it. And at this point East and West have met. With two short walls already up, the bracing's in place for the long walls, and once you've read about framing them Western style, any style-to-

A long wall, short wall, and bracing

style changes will be self-evident. Try it and you'll see what I mean. You've already built a deck and a couple of walls by now. Extemporizing in wood is becoming second nature.

Doubling the Plates

Finally, to nail down the uppermost top plates, find an A-frame ladder by whatever means one generally finds ladders. A six-footer should do nicely. Then array each 2x4 so that it's centered atop the lowermost top plate beneath it. If it's in two pieces, make sure the ends are butted smartly together. Now, using 16d nails, nail it down into the ends of the studs, one nail per stud, two nails at either end of the board. Be sure it's lined up well with the lower top plate. The $1\frac{1}{2}''$ sides of both 2x4s should form a single surface, and if they're crowned very differently this may take some jiggling. The best way to compensate for mismatched crowning is to put the nails in in sequence (as opposed to nailing both ends and then attending to the middle). It gives you the most leverage if you have to bend the board a little before you nail it. And if you do have to bend it and there's no one around to hold it while it's bent, first grab it by the end, bend it, put a nail in anywhere that will hold the bend, then drive in the *real* nail (or nails), and finally, after the real nails are holding it in place, remove the temporary nail. In fact, get used to inventing all kinds of mini-techniques to accomplish the results you're after. Framing a building continually calls for feats of improvisation, feats which may be meaningless to anyone who isn't framing a building but which sometimes give more private satisfaction than the carpentry itself.

Squaring Up the Building

Whichever way you build, Western-style or Eastern, there's one practice you should religiously observe. After you finish a wall—any wall—square it up, brace it diagonally, and brace it well. Of course you'll try to keep it square while you're building it, but there's always some touch-up squaring to be done when it's finished, and even a fully framed wall has some play in it. So, after the last nail goes in and before it's connected to any of its neighbors, use the equal-diagonal method to ascertain rectangularity and diagonal braces to *retain* rectangularity. Frame structures are crazy for true right angles. The more you give them, the happier they are.

Top plate butt joint should fall above header whenever possible

7

Framing the Roof

Deciding on Pitch and Overhang

Of all the tasks that building the outbuilding calls for, roof framing is unique in that square corners take a clear backseat to more exotic ones. Therefore, not only does framing the roof carry you literally to the high point of the project, it also gives relief from the surfeit of right angles that you're undoubtedly experiencing. It's true that 90° angles are a tad easier to cut than ones of, say, $37\frac{1}{2}°$, but on no account should that be cause for much anxiety. Guys whose evolutionary disadvantage would be apparent even at a convention of Cro-Magnons have gotten the hang of rafter cutting, and the chances are excellent that you can get it too.

Before you cut anything, though, decide how much to pitch your roof. The roof of the outbuilding shown in these pages is pitched 6″ to the foot (i.e., it rises 6″ for every horizontal foot it spans), which is about as steep a roof as you can expect to stand up on without risking extensive bodily damage in the best of circumstances and outright death in the worst. That's not to say you shouldn't pitch your roof more steeply. You may be building in the far north, where snow loads get so heavy that in order to unburden themselves, roofs need to take occasional advantage of the avalanche effect. If that's so, get some local advice on proper roof pitch or simply measure the pitch of a building that's shown its durability over the years. Then again, sheer aesthetics may demand that you pitch the roof of your outbuilding to match that of some larger structure nearby. If the pitch is steep, all it means is that you'll have to do some temporary carpentry. With sufficient scaffolding you can comfortably clamber around on just about any roof at all.

Of course, you can pitch your roof at less than 6″ per foot too, particularly if your climate is a mild one. To my taste, roofs pitched at less than 4″ make the structure they're topping off look squat and boxy, but they're certainly a cinch to walk around on. Probably the best thing to do is look at houses, see what pleases you, and then, with the understanding that a steep roof will call for extra scaffolds and therefore extra labor (not a whole lot) and extra lumber (though once you dismantle the scaffolds the lumber is perfectly reusable), choose what you like the most.

When you know what you want, examine how a rafter goes into a frame structure and also examine how a rafter is cut.

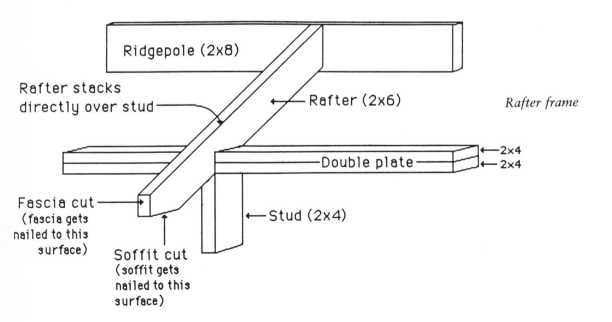

Rafter frame

Whole rafter

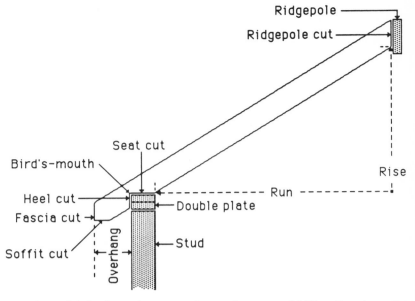

Now think about how much overhang you'd like. Don't make it less than 6″ (you'll need that much room to install soffit vents, about which more later), and don't opt for so much overhang that people entering the outbuilding are likely to skin their skulls on the fascia. Let your prevailing climate influence your decision. Less overhang keeps the windows from being shaded, which gives you more interior light and also more heat. More overhang keeps the outsides of the walls and windows dryer when it rains and shadier when the sun is out, but too much can make the inside of a building seem drearily dark. Overhang tables have been compiled and published in a number of construction books, but most of them correlate overhang with passive solar heating efficiency, not quality of life. The best way to make your choice is to look at existing buildings nearby and select what suits you.

That done, you're ready to mark your master rafter, the template whose outlines you'll trace onto all the others. If you don't have a 2x6 to scribe the marks on, it's probably because you haven't ordered any, which, in turn, is probably because you'd like to know how long they ought to be before you order them. Fair enough. But first an observation in passing.

A Full-Size Drawing

There's a tool called a rafter square that makes drawing lines for all the cuts—the heel cut and seat cut of the bird's-mouth,

Marking rafters

the fascia and soffit cuts at the lower end of the rafter, and the ridgepole cut at the upper end—easier than they'd be to draw without a rafter square. However, unless you're planning to build a *number* of roofs, I'd like to propose an approach that's more straightforward and certainly more field-expedient. On the subfloor of your deck, simply draw a full-size diagram of your roof, like so . . .

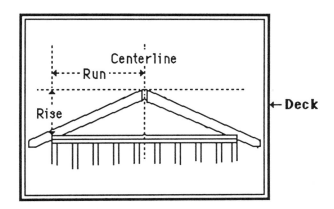

Your diagram (drawing the studs isn't necessary)

Full-scale drawings are admittedly unconventional, but this one will spare you from having to master the intricacies of a rafter square, and it will also protect you against errors of scale. With a rafter square you'd be working basically in inches, while when you frame your roof you're working essentially in feet.

So, roughly speaking, you'd be making your rafter marks at $\frac{1}{12}$th scale. Which is to say that if you want your roof to rise 4′ for every 8′ of run, you'd translate that into something like 4 *inches* of rise for every 8 *inches* of run when you mark the rafters. Now, that's fine on paper. The trouble is, working at $\frac{1}{12}$th scale gives you every opportunity to multiply any piddling little error you make by twelve. Your sawblade makes a $\frac{1}{8}''$ cut whether you're working at $\frac{1}{12}$th scale or full-scale, and an error of $\frac{1}{8}''$ at the ridgepole cut of a rafter can easily produce an error of 1″ or more at the other end, when you're actually assembling the roof and wondering why the seat cut doesn't sit anywhere nearly flush onto the wall plate.

Ordering Lumber

Also, from the full-scale drawing you can easily derive the minimum length of the 2x6s you'll be making rafters from, and you can even play a little first. For instance, if it turns out that your choice of pitch and overhang is such that you need 8′2″ 2x6s for your rafters, which mandates that you buy ten-footers, you may decide to shave a couple of inches of overhang and make do with *eight*-footers, which would reduce waste and consequently save some money. Or, as yet another labor saver, you might want to adjust the overhang so that the soffits go in just as they come from the lumberyard, without your having to rip them down to size. To do that, read ahead to the chapters on closing the roof and sheathing and siding and also look at the diagram.

If you decide right now what kind of sheathing/siding and soffit boards you'll use and then find out in what widths the soffit boards are available, you can tune the overhang so that the horizontal distance between the inside of the fascia and the outside of the sheathing/siding is exactly equal to the width of the soffit board(s), so the soffits can come right off the truck and into the roof. (All right: you still may—I said *may*—have to cut them to length.) Whatever you do, make sure that the seat cut of the rafter is a full $3\frac{1}{2}''$ long, so it covers the whole top surface of the wall plate. The heel cut can fall as it may.

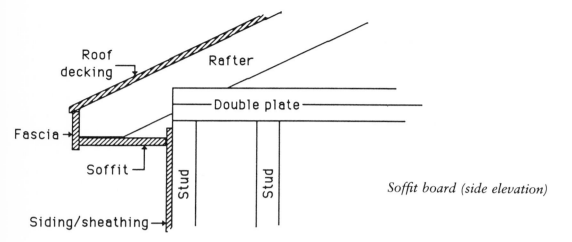

Soffit board (side elevation)

You're ready now to order the 2x6s destined to become rafters (twice as many rafters as you have joists) and the 2x8 that will be your ridgepole. The ridgepole runs the full length of the outbuilding, of course, and if you can't get it in one piece, make the two pieces long enough so that you have at least an extra foot or so of play. Also, order enough extra 2x6 so that you can make two $14\frac{1}{2}''$-long scabs. (That doesn't mean you have to buy a 29" length of 2x6; you can add some extra length onto one or two of the rafter 2x6s and cut the scabs from those.) And order another pair of 2x6s, each one as long as the width of the building. Ultimately you'll use them as tie beams (to keep the roof from prying the long walls apart), but even before that you'll need them for some temporary carpentry. Here, as with the studs and plates, my preference is for Douglas fir, but virtually any wood that's hospitable to nails will do the job.

Cutting Rafters

When the lumber truck arrives, find a particularly clean and uncrowned 2x6 and designate it as the template rafter. Lay it over your full-size deck drawing, mark all the cuts, and then cut it as precisely as you can. Now set it back down onto the drawing. If it pretty well conforms, that's your template. If not, save it for use as a rafter (by now you have your own feeling for what tolerances are allowable), but make another template so you don't carry the error over into all the other rafters. When you have a template you're satisfied with, scribe its outlines onto the other 2x6s and cut them. Unlike some of the other repetitive jobs you've done, which had a pleasing, even a euphoric rhythm to them (especially when each repetition gave you a palpable new increment of proficiency with a saw or a hammer or just your hands), cutting rafters may feel like the sort of assembly-line work that the Japanese reserve for robots. You're impatient to see your roof up, and your impulse may be to do a quick and dirty job with the rafters. Don't. Rafters that seat properly make roof framing an interesting and gratifying task. Rafters that seat poorly leave you with a big, unwieldy mess. So check all the bird's-mouths, and, wherever necessary, touch them up with a handsaw.

Now prepare your ridgepole. If it's in one piece, cut it to the precise length of the outbuilding and mark where the rafters will butt it on both sides. (The butt joints will fall directly over the joists, so you can copy your old joist marks onto the ridgepole.) If it's in two pieces cut them so the seam falls halfway between a pair of rafters. Decide which edge of the ridgepole is the top and nail on a pair of $14\frac{1}{2}''$, 2x6 scabs. Place the scabs between the rafter marks and set their lower edges flush with the lower edge of the ridgepole . . .

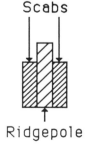

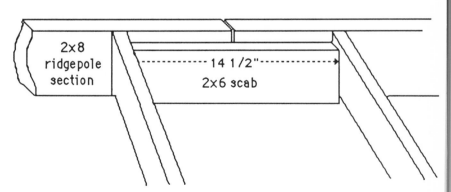

Nail the scabs securely to the ridgepole. You want the two lengths of 2x8 to be as one. And while we're talking about doing things well, this is a good time to make sure your wall frames are well braced. You're going to be climbing on them, and the more secure you feel the better your carpentry's likely to be. Long boards make the best temporary diagonal braces, but good-sized plywood fragments work well too—whatever it takes to keep the frame from tilting under stress.

Raising the Ridgepole

And, although it's possible to put the ridgepole up all by yourself, I strongly recommend that you get help. There's simply too much chance of lumber coming down on top of your head if you work alone. However, you can get *started* by yourself by fashioning a couple of A-frames like so . . .

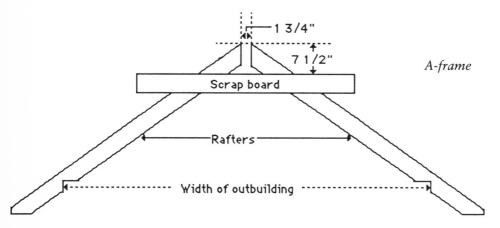

The A-frames will be heavy, but with someone down on the deck to hand them up to you, you shouldn't have that much trouble lifting them up onto the plates. Incidentally, nail the scrap board across the two legs fairly well (without driving the nails in all the way; you'll eventually pull them off) and measure the distances carefully. You want the A-frame to be rigid, and you want it to fit.

Now stand the first A-frame in place, up over one of the ends of the outbuilding. Your ridgepole, which still isn't attached to anything, can be useful here. If your assistant helps you slide an end of it into the slot at the top of the A-frame and leaves its other end resting down on the deck, it can function as a third leg and hold the A-frame standing temporarily. Given that your wall frames aren't exactly machined to close tolerances it's unlikely that both bird's-mouths will seat perfectly onto the plates. But start with either one, seat it well (even if it means teasing the other bird's-mouth into place with a hammer), make sure the rafter is flush with the edge of the building, and toenail it down to the plate (a pair of 10d nails on either side—and be careful with the hammer: bird's-mouths are cut obliquely to the grain of the wood and therefore splinter easily). Starting now, you'll be moving your ladder around a great deal and climbing up and down a great deal. It's par for the course, so you may as well get used to it. Always place it so you can hold up whatever weight you need to bear without being thrown off balance. This will require a bit of thought at first, but in short order your body will learn where the ladder wants to be.

As for that second bird's-mouth—the one that's unlikely to seat perfectly onto the plate—don't worry about fitting it too carefully just yet. All you need to do right now is tack it temporarily in place. (Of course, if it *does* seat nicely, seize the day and nail it permanently.)

Next find a long board—any long board—and tack one end of it to the upper part of either leg of the A-frame. When you've done that you can use the board to adjust the A-frame until it's vertical. (You can be up on a ladder, testing the A-frame with your level while your confederate is down on the deck manipulating the board.) When the A-frame *is* vertical, brace everything in place by tacking the other end of the board to the deck. Now repeat the whole procedure for the second A-frame, and when you're finished you'll have a pair of slots, one at each end of the building, to hold up the ridgepole.

What you have to do next is get the ridgepole into the slots. One way to do it is to take one of your tie beams and tack it on its flat side across the tops of the walls. Keep it close to one of the ends of the structure—it doesn't matter which one—so that it forms a kind of shelf. Now, with help from your assistant, lift the ridgepole up onto the plates so that it spans the length of the building and one of its ends is sitting on the shelf. What you want to be able to do is place one end of the ridgepole into

End rafters and ridgepole

an A-frame slot without having the other end slide off the plate, fall down onto the deck, and dent the unsullied perfection of your subfloor. If you haven't positioned your 2x6 "shelf" in a way that lets you do that, reposition it and try again. When you've got an end of the ridgepole seated in an A-frame slot, nail it in place with a lone 16d nail. (Once again, single nails permit rotation.) Now you can move your ladder back to the other A-frame, lift the free end of the ridgepole up off the "shelf," fit it into the second slot, and nail it with just a single nail. By hook and by crook, you've gotten your ridgepole up.

Installing the Rafters

The rest is easy. Before nailing the ridgepole to the A-frames with any finality, nail on all the other rafters. Make sure they're stacked directly over their corresponding studs, that they're on their marks, and that their uppermost points don't come up any higher than the top edge of the ridgepole. For each one, face-nail a couple of 16d nails down through the end of the rafter into the ridgepole, then toenail a few 10d nails through the *side* of the rafter into the ridgepole. At the bird's-mouths, toenail a couple of 10d nails through each side of the rafter near the seat cut (trying to nail with the grain of the wood, so you don't splinter it) and down into the wall plate. To keep things in balance, nail the rafters in opposing pairs. And, after every couple of pairs, check to make sure the roof isn't becoming lopsided.

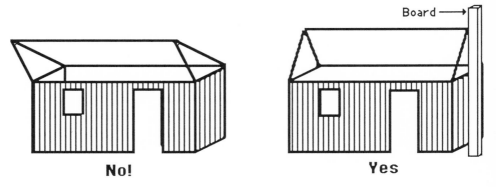

Line up a long, straight board with the outside of one of the end walls and touch it to the end of the ridgepole. If the ridgepole end isn't plumb with the wall, find something heavy and bang it hard until it is.

Installing the Tie Beams

Another thing to do after you've got a couple of rafters well nailed is to install the tie beams. Cut them to the exact width of the outbuilding and trim off a corner from each end so the edge of the tie beam is flush with the tops of the rafters. (You can use the full-size diagram on the deck to mark the cuts.) Then determine which pair of rafters is most nearly a third of the distance between the ends of the building and which pair is most nearly *two* thirds. Let's call them pair one and pair two. Place a tie beam on edge across the top plates so its flat side will come right up against the flat sides of rafters in pair one. Then toenail it down to the top plate. Now do the same for the second tie beam, this time making sure that its flat side will come up tight against the flat sides of the rafters in pair two. When you put in the rafters of pairs one and two, face-nail them to the tie beams (or face-nail the tie beams to the rafters). And don't wait to do all this until you've already put most of the rafters in. By that time the roof may be spreading the tops of the two long walls far enough apart that you can't nail the tie beams where they're supposed to go.

Once the ridgepole is properly supported by real rafters appropriately nailed, you can pry the scraps off the A-frames at the gable ends and nail the end rafters just like the middle ones. Then you can come down off your ladder and see what's been accomplished. You've framed the roof.

Framing the Gable Ends

But don't rest on your laurels *too* long. You still have to frame out the gable end walls with 2x4s, leaving ROs for any windows you'll be putting in. By now you know how to do this, so you don't need much advice from me. Just make sure that the studs in the gable ends stack directly over the studs in the end walls, that they run straight up and down, and that you cut their tops at an angle which conforms to the pitch of the roof. Then just toenail them with three or four 10d nails at either end.

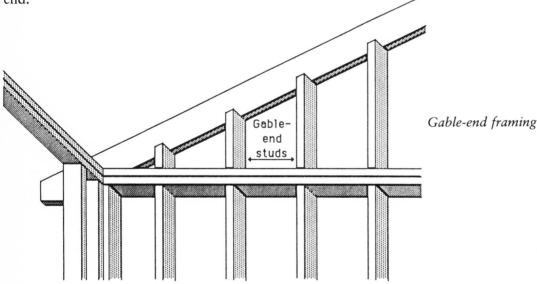

Gable-end framing

Gable Nailers

You'll finish off the job by installing four gable nailers—2x4s that run flat along the tops of the gable end studs. (Since only $1\frac{1}{2}''$ of each gable-end stud top is occupied by rafter, the remaining $2''$ is available for end nailing.) Why are you being asked to cram your aching body into all manner of splintery crevices to do this? Because when you want to nail some sort of ceiling to the undersides of the rafters, you'll wish that someone had had the foresight to arrange that there be something to nail the outer edges of the ceiling to. Inevitably, dear reader, that someone is you.

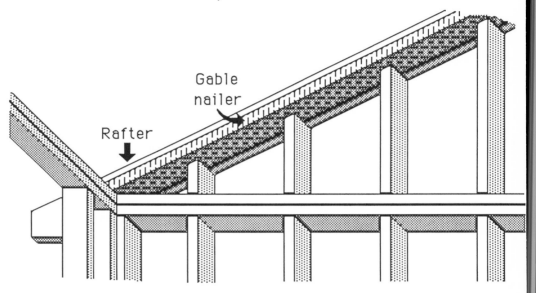

8

Sheathing, Siding, Windows, and Door

THE EXTERIOR WALLS

For centuries, virtually all frame dwellings were sheathed and sided separately. The sheathing usually consisted of tightly butted boards nailed diagonally across the outside of the frame. The siding was clapboards, shingles, barn boards, stucco, or any number of other materials that were applied over the sheathing. The sheathing braced the frame (which is why it's a good idea to nail it on before you put up the roof) and also served as a rudimentary enclosure. The siding sealed out the weather and presented an agreeable facade.

Nowadays some frame structures are sheathed and sided separately and some aren't. When they are, the sheathing is almost always plywood or chipboard (a composition of wood chips and assorted resins that bind the chips together), which gives a thinner skin than old-fashioned board sheathing. Nonetheless, the laminations in plywood make for great strength per unit thickness, and the fact that there are fewer seams makes for a good deal less air infiltration. And, of course, plywood and chipboard don't require as much labor to put up. Normally, builder's felt (the kind that's saturated with asphalt) is tacked over the sheathing, then the doors and windows are nailed in place, and finally the siding is put on in a way that seals the outside edges of the doors and windows.

There is, however, a shortcut. Let's suppose that someone not unlike yourself, having just framed out a small building, is hot to get it done and move on in. The framing already defines the building so thoroughly that the remaining work seems like mere bookkeeping. He therefore wants it to go as fast as possible—not at the expense of quality, but not at the expense of great amounts of time or money either.

To his great good fortune, his local lumberyard can supply him with a material that satisfies his requirements. I'm speaking of combination sheathing/siding, plywood made to do both jobs at once, the best-known variety of which is T-111—"texture one-eleven." T-111 and its clones, which come in 4x8 sheets that are either $\frac{3}{8}''$ or $\frac{5}{8}''$ thick, is plywood sheathing with the siding glued right on. (The siding is the outermost ply.) It can be obtained with its "good" side smooth or rough-sawn and with lines inscribed to make it look like boards. The lines can be vertical or horizonal, and far apart (wide ersatz boards) or close together (narrow ersatz boards). It's easy to put up, it braces your frame, it takes stain or paint or clear preservative more readily than just about any other species of siding, it weathers to a pleasing gray if left unfinished (at which point it can be coated with exterior wood preservative), it's inexpensive, and—since by now you're aching for completion—it gives you early closure. Moreover, if you put it up and then don't like the way it looks, you can nail shingles or clapboard or cedar tongue-and-groove or some other kind of siding right on top of it without its giving you the barest squeak of protest. And meanwhile you'll still have a weathertight, presentable, functioning outbuilding.

I therefore strongly recommend you go with T-111 or a look-alike. Visit your lumberyard and see what kinds they stock. If you don't care for what they have, visit the competition or browse through a catalogue and order what you want. The $\frac{5}{8}''$ is more solid and allows the board lines to be notched more deeply, which makes it look a trifle more convincing. The $\frac{3}{8}''$ drums are a little louder during horizontal hailstorms, but it's more than strong enough for a one-story outbuilding, and being lighter, it's substantially easier to handle if you're working alone.

The stuff is graded by quality, i.e., by how many of those football-shaped plugs have been glued into the outer ply to replace knotholes. Examine a few sheets of different quality, see how many "footballs" per square yard you can tolerate, and, once you've made your mind up, order one sheet for every 4' of exterior wall plus enough extra to close in the gable ends. And when you place the order, include a few pounds of galvanized 6d nails, a box of $\frac{3}{8}''$ staples, and some rolls of 15-pound builder's felt. Each standard roll is 3' wide and about 144' long. Figure about 15' of felt per sheet of siding. We'll go into more detail presently.

Installing the Door and Windows

While you're waiting for the siding to arrive, you can install the door and windows. So get a roll of 8" aluminum flashing if you don't have one already, and also get a roll of plastic window flashing. Since the siding will seal up all the window-to-wall-frame interfaces, it will also seal out moving air and sunshine, which ordinarily keep wood joints dry and relatively free of rot. Therefore, certain rot-prone crannies need to be flashed, and the frames surrounding door and window units are prominent among them.

Checking the ROs

However, don't start flashing until you try the door and windows in their corresponding rough openings. The ROs have to be big enough to accommodate them and small enough to let you nail them in. (You nail right through the outer side casing—the part that's like a picture frame—into the members that make up the RO.) If the ROs are too big, build them up with extra boards. If they're too small, woe betide you. You have to take them apart and rebuild them to size. Unless, that is, they're just a teeny bit too small, in which case you may be able to shave them down—or is it *up*?—with a plane. If that's what you elect to do, though, make sure not to plane any 2x4s all the way down to 1x4s and then wink at the shortfall. Ineluctably, your commitment to the outbuilding has become so powerful by now that its structural integrity and your own personal integrity have become one and the same thing. The building code has Mosaic resonance in the universe of construction carpentry; when you flout it blatantly, you bring shame and wrath on novice builders everywhere (not to mention the mysterious misfortunes—ones that often have nothing to do with building—which start befalling you).

Flashing, Leveling, and the Door and Windows

Back to work. Once the ROs check out, tack plastic flashing around them, and cut pieces of aluminum flashing for where the bottoms of the window casings sit on the sills. For the door RO, line the entire bottom with metal flashing, and do it so some of the flashing will extend past the door casing both inside and outside. You can trim the overage later. Then, one at a time, get the windows into their rough openings, level them, and nail them. You level a window by putting your level across one of the unit's horizontal members and inserting shingle shims between the bottom of its casing and the sill of the rough opening until the level says your window is level. This procedure inevitably turns out to be necessary: the sills of your rough openings may be level*ish*, but it's too much to expect them to be flawlessly level. (On the other hand, I wouldn't level the door if I were

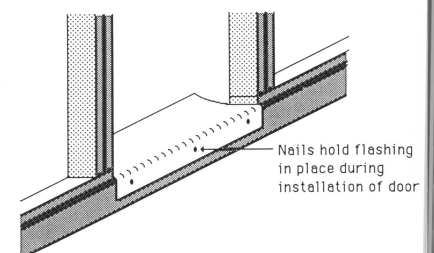

Bottom flashing for door

you unless the subfloor is grossly out of kilter, which it oughtn't be in the first place.) After each window is level, use galvanized nails to fasten it in place. As for how many to use, rule of thumb dictates that few are better than lots. You'll caulk the windows later, which will serve to glue them in, and you'll trim them from inside, which will fasten them still more. Their outer side casings present cheerful, smoothly milled faces to the world, and the last thing you want to be called to account for is pocking them too abundantly with nailheads.

(A caveat: door and window side casings are most commonly made of pine, which is particularly prone to splitting when nails are driven through it at an angle. Yet it's sometimes necessary to do just that, especially where your rough opening is on the large side and the outer casing just barely laps over the RO members. If you do have to go in at an angle, try using long Sheetrock screws instead of nails. Predrill the casings at the angle you want—the flimsiest of drills will suffice—and then put the screws through the holes and into the RO members. This takes a mite more time than nailing, but the casings will have a better chance of remaining intact.)

(Another caveat: some of the temporary bracing on your walls probably runs across your rough openings. When you remove it to put in windows or the door, remove only as much as you need to, and then promptly find some other way to rebrace the frame. As rigid as they may appear to be, window units are not braces. And when you do rebrace, do it from the inside of the wall, so you won't have to remove the braces yet again as you nail on the siding.)

Now, to keep rain from trickling into the rough openings, make strips of top flashing for the windows and door. For each unit, cut a strip of 8" aluminum that's a hair longer than the width of the outer casing and, using a board as a kind of anvil, bend it so it will eventually fit under the siding and over the casing. Then tack each strip of flashing over its unit. A couple of tacks through the flashing and into the header are enough. All you're really doing is holding the strips in place until you nail the siding on. Which you're finally ready to do.

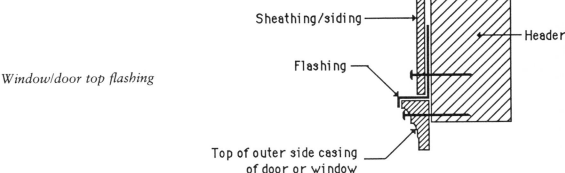

Window/door top flashing

INSTALLING SIDING

Start by picking any stud in either of the long walls. Fit your carpenter's square up against the soffit cut of the rafter just above it, and mark the height of the soffit cut on the stud. Any siding above that point won't be exposed to weather, since that's where the soffit will meet the siding. Now pick a point a couple of inches above that point, and measure 8' down. That's where the bottom edge of the siding will be. Does it please your eye? Yes? Good. Leave it alone. No? Play with it some more. Just don't let the *upper* edge come down below the soffit, and make sure the outer wall will be sided to a point at least 1" below the underside of the subfloor.

Temporary Ledge

Now once again measure 8' down from the first point on the stud. The 8' mark should fall somewhere on your boxing. Nail a good long board along the boxing to serve as a ledge for sheets of siding. Don't nail it especially well, but un-crown it as you nail. You want the ledge to be as straight as possible. Test it by placing a sheet of siding on it and then seeing if the vertical edges of the siding are parallel to the studs. Typical studs are anything but straight (a statement that could be inflammatory in other contexts), but your mind's eye can furnish you with an average straight line down through their centers. Try to extend the ledge along the full length of the wall. It needn't be made up of a single board, nor do the boards even have to be of the same kind. Once you nail the siding on you'll pull the ledge off anyway. All that's important is that its upper edge be a straight horizontal line.

Temporary ledge to support siding

Notice how the vertical edges of the T-111 are milled. Each sheet laps over its neighbor. Keep this in mind as you put the siding up. Don't be caught having to lap any sheets *under* a sheet that's already nailed. Put the stuff up so that the bottom lap of every lap joint is at the leading edge.

Siding

Where sheets have to be cut, do it so the siding butts neatly up against the edges of the outer casings of whatever doors and windows are in the way. The measuring is a little tedious because it has to be precise. But it's also worth the trouble. Yes, you can fill any yawning seams between the casings and your cuts with caulk (the seams have to be caulked anyway, even if they're perfect), but caulking is intended as a sealer, not a substitute for siding. To make the cuts, you may want to treat your circular saw to the luxury of a plywood blade, whose baby teeth will splinter the T-111 appreciably less than the dinosaur dentition of the all-purpose blade you've probably got in it now. Either way, adjust the saw to cut shallowly. The siding will reward you by looking prettier.

Builder's Felt

When you know you'll be installing insulation and/or interior walls, you'll also want to line the inside of each sheet with

Cutting the siding

builder's felt before you nail it. Three horizontal courses per sheet will be sufficient. Start at the top of each sheet, let the felt extend a couple of inches over the top edge, and lap each course a few inches over the next, stapling as you go. You don't need any more staples than it takes to hold the felt in place (and if you're using $\frac{3}{8}''$ siding, don't use $\frac{5}{8}''$ staples!), and you might try to line up the edge of the bottom course with the lower edge of the siding. That way you won't have to trim it later. Let each course extend a few inches past the leading edge of the T-111 sheet (by "leading edge" I mean the vertical edge that will be lapped under the next sheet to come) to ensure that there's felt under the lap joint.

Installing the siding

Tacking felt directly to the stud wall

If there's no wind to tear the felt apart, an even faster method is to staple it directly to the frame. Tack the first course so the bottom of the felt falls a little below the edge of the subfloor and, unrolling the felt as you go along, put in enough staples to hold it neatly in place. Each subsequent course should lap a few inches over the one just below, and you don't have to cut holes for the windows and doors until you're in the mood. When you're done, you'll have an authentic tar paper shack that's perilously vulnerable to stiff breezes. So either nail on all the siding before the wind comes up or, more practically, don't put the felt on all at once. Tack up only as much as you can cover with the sheets of T-111 on a given day.

You'll probably want to skip using felt if you're certain you'll soon be shingling over the T-111 or if you plan neither to insulate nor to put up any interior walls. When you're shingling, you staple felt *over* the siding, then shingle over the felt. And if you're not installing Sheetrock or paneling inside (which is a perfectly reasonable thing not to do) you'll find that the inside surface of the T-111 makes far more attractive (and less smelly) walls than tar paper. Do caulk the seams between the sheets extra carefully, though. Later down the road, if you decide to insulate after all, you can use felt-faced insulation.

While you're putting up the siding, don't be too concerned with the edges at the corners of the building. Concentrate instead on making sure that the lap joint at the leading edge of the first sheet on every wall falls over the center of a stud. In due course the corners of the building will be covered with corner boards which are broad enough to conceal some pretty gross discrepancies.

Caulking and Nailing

Caulk the lap joints with a thin bead of exterior-quality latex caulk before you nail them. If you plan on leaving the siding its natural color, gray caulk will ultimately be less conspicuous than woody brown. If you plan to paint, the color of the caulking doesn't really matter.

You may remember that not so long ago we spoke about making sure that the studs at every 4' point on a wall weren't too badly crowned. Now you see why, yes? As you nail up the sheets (use galvanized 6d nails about 1' apart along every stud), nail right through each of the lap joints. That way you'll fasten the edges of both lapped sheets at once.

Flashing and Siding the Gable-Ends

After you've got the first 8' of siding up all around the building, you'll have to cut some shaped pieces to close the gable-ends. Do it so their vertical edges line up with the vertical edges of the rectangular pieces beneath them. If the cuts just under the roof edge aren't perfect, don't fret; you'll cover the seams with fascia trim. And don't nail the gable-end siding on until you've flashed the upper edge of the bottom 8' course on each of the two "short" walls. Otherwise rainwater streaming down the gable-end siding can be blown through the horizontal siding seam and into the walls. (The way T-111 is made, it gives you caulkable lap joints for your vertical seams but not your horizontal ones.) You do this in much the same way that you flashed the tops of your door and window casings . . .

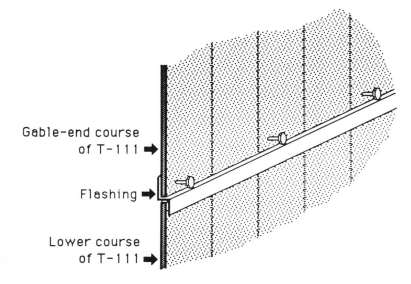

Flashing the T-111 on the short walls

You can make the flashing yourself, which can look a mite on the rough-hewn side, or you can buy flashing that's preformed for the purpose. The latter usually comes painted a dull brown, is cheap, and tends to look considerably neater. If you do decide to buy it, and if you find that the only flashing in stock is for $\frac{5}{8}''$ siding even though it's $\frac{3}{8}''$ you're using (a vivid possibility), use it anyway. The vastness of your exterior walls will eat up the mismatch.

With all your siding up, the wall frames are as braced as they'll get, so you can knock/pull/pry off all the temporaries. However, do remember that your windows are in. Glass hates flying 2x4s.

Caulking Again

Once the siding's all up, caulk around the outside of the door and windows. (You don't have to caulk the flashed tops.) You're working now with—how shall we say?—facade, so do it carefully. Fine motor control can be difficult to summon when you're called upon to make an abrupt transition from the gorilla crudity of hammering nails through big pieces of wood to the finesse of smoothing latex caulk. But making the transition is well worthwhile. Let the breadth of spirit for which you're justly renowned furnish you with the resources you need. And if you're too impatient now to fuss that much with caulk, be consoled in the knowledge that miracles of concealment can be achieved with paint and store-bought molding.

I'd say that this is an appropriate time to stand back and take the measure of what you've accomplished. You have a building now—complete with walls, door, windows, even an openwork lattice roof that washes the interior with sunlight. Maybe there's a part of you that wants to say, "Enough," a part of you that's getting bored, disaffected, ready to create an agenda of fresh preoccupations. I think you know the part I mean: that creature inside everyone who rebels against completion, who comes fully alive about two thirds of the way through most large projects and begins to mutter, "The hell with all this stuff. What's next?" Please understand: it's not my intention to criticize that part of you in any way. Completion problems are as endemic in the building trade as lung disease is in asbestos work. In fact, you may listlessly decide that this is as far as you want to go. Even if the outbuilding isn't yet a studio or office or retreat or whatever it's supposed to be, it's ready to do yeoman duty as an imposing monument to your legendary powers of procrastination. You may already have what you need.

9

Closing the Roof

Just so we all know where we're going, let's attend to procedural matters first. This, very quickly, is what you'll do to close the roof:

1. Nail $\frac{1}{2}''$ plywood decking across the tops of the rafters.
2. Nail metal drip edge along the outer borders of the decking to keep the weather from penetrating the edges of the plywood (optional).
3. Nail a layer of roofing felt over the decking.
4. Nail the roofing of your choice over the roofing felt.
5. Install a ridge vent along the full length of your roof ridge.
6. Vent and install the soffits.
7. Nail on the fascia.
8. Nail the corner boards onto the corners of the outer walls.
9. Finish up the roof trim.

You'll notice that most of the roof closing could have taken place before the siding went on, but in strong winds closed roofs can behave like airfoils and stress their underpinnings in sudden, unpredictable ways. Sheathing/siding gives the walls the rigidity they need.

WORKING UP ON THE ROOF

Before we go ahead and put a few layers of skin over the airfoil, however, there's the question of how in hell you're supposed to get yourself and your materials safely up onto the roof. Because with the experience you've already accumulated, most of the rest will be self-evident.

Now, since I don't know how much you've pitched your roof, I can't furnish you with any prefabricated solutions to the problem of negotiating the rooftop. Still, the key to it all is temporary carpentry: scaffolds, platforms, skids, shelves, whatever does the job. You can gripe about doing so much sawing and nailing and (if circumstances don't permit fruitful scavenging) wood-buying only to have to pull it all apart when you're done, or you can relish the chance to do some improvisatory, freeform construction work—work in which appearance counts for nothing and safety, utility, and economy for all.

Skids for Raising Your Materials

Probably the hardest part of closing the roof is nailing the bottom course of decking on each side. Your outbuilding is modular, so you may not have to cut any plywood until you're doing the second course, by which time the edge of the *first* course can hold the sheets in place as you nail them. But those first-course sheets have to be gotten up to the roof and then secured so that they don't slide down and slice you into ragged halves. To do it, get someone who's fairly strong to help you and then try using long boards for skids. Nails driven partway into the top surfaces of the skids will keep the plywood sheets from sliding downward. Climb up onto the roof and find a place where the footing feels secure. (A ladder isn't such a place; it's much safer to stand on the framing of the outbuilding itself.) If you can't find a good foothold, make one by nailing some scrap boards to the frame. You can pry them off when you no longer need them. Now have your helper slide a sheet up to you along the skids. (If you've concluded that this isn't something to do on a windy day, you're right.) Once you can grab the top of the sheet, tell your helper to get out from underneath it. Then pull it up onto the roof and put a nail through the sheet and into a rafter.

Incidentally, you don't have to pull the sheet all the way up onto the roof before you put in the temporary nail. If you nail the top of a sheet to a place somewhere near the bottom of a rafter, you can then rotate the whole sheet on the nail so that it climbs further up onto the roof. When the sheet is sitting up on the rafters in what seems like a good place, put in a second temporary nail through the other end and into the roof frame. That will ensure against a gust of wind lifting the whole sheet (a real possibility) and hurling it where it might injure someone. Now repeat the procedure until you've hoisted enough plywood to do a whole course of roof decking.

Roof decking on skids

Temporary Slide Stops, Ladder Safety

Once your plywood is up, you have to provide a means of keeping it from sliding down the rafters while you're working with it. One way is to rig up some temporary slide stops. They offer the advantage of serving two purposes at once: they keep plywood (and feet) up on the roof, and they provide a ledge to lean the ladder against. And try to form good ladder habits. One I recommend is that, every time you climb, you make a point of remembering to bring whatever tools you'll need. The

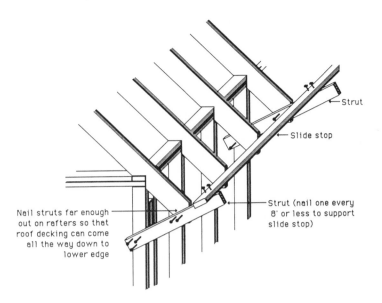

Slide stop

more extra trips you make, the more frustrated you'll become, and a happy climber tends to be less accident-prone than a disgruntled one. Another habit I recommend is that of tethering your ladder to the roof edge. It's embarrassing and worse to take that last crucial step onto the roof only to discover that you've kicked the ladder out from underfoot and thereby burned all bridges between you and terra firma.

The slide stop is only one kind of scaffolding, but, whatever kind you build, take pains to observe the guidelines in the diagram on the facing page. Scaffolding should always enhance your safety, never imperil it.

After you've got the two bottom courses of plywood on, everything becomes easier. You can make all kinds of interesting shelves, steps, and ledges and simply nail them to the plywood. Never mind that you'll dismantle them. Sometimes means *are* as gratifying as ends. Use some of what you've learned and be ingenious. (Notice that I didn't say "creative," a repellent word on which there ought to be a ten-year moratorium. Whatever it once meant, creativity has come to be a synonym for mediocrity run rampant. So do your fellow earthlings a favor and don't contribute to the clutter of "creative" geegaws that are already choking the planet half to death.) Incidentally, in this

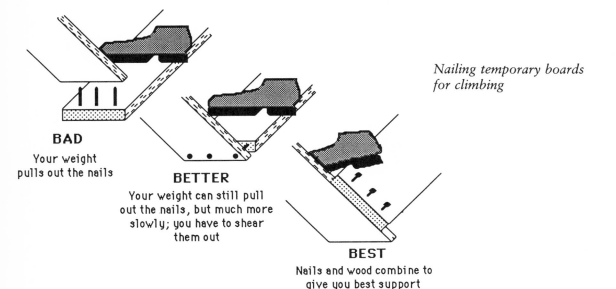

Nailing temporary boards for climbing

flowering of your ingenuity you needn't worry about nails puncturing the roof decking. The decking's there to hold up the roofing (and to keep the rafters rigid), but it's not supposed to *be* roofing.

Of course, all this presupposes that you're not afraid of heights. The superstructure of the outbuilding can bear your weight quite comfortably, but your psyche may not be as supportive as the rafters. If heights bother you, the worst mistake you can make is to pretend they don't. Counterphobic behavior sometimes has its places, but the surface of a roof isn't among them. I've talked to several builders who claim to have deconditioned themselves against fear of heights, and all agree that if you find yourself feeling anxious, it's not a good idea to challenge the anxiety. The fear usually comes in fits and starts, and the best thing to do while you're feeling it is to stop whatever you're doing, give yourself a solid foothold and let the worst part pass. And if even that doesn't work, tap into your reservoir of loyal friends and relatives, especially those who are poised to jump on any opportunity to feel superior to you. When you confess to your phobia, they'll be more than happy to scale your parapets and nail on your roof. Probably they'll even take direction from the ground—anything that lets them say, every three minutes or so, "See? There's nothing to it." In fact, if you're in a builder's lull around now, you may want to *pretend* to acrophobia even if you don't really have it. The benefits may far outweigh the needling.

Now to the roof itself.

INSTALLING THE ROOF

Decking

You nail on $\frac{1}{2}''$ exterior plywood (no need for any higher grade than CD) in the customary alternating pattern, with the long edges of the plywood at right angles to the rafters. Use 8d nails, one along every foot of rafter. Lap the edge of the lowest course about $\frac{3}{4}''$ over the rafter ends (so the fascia can fit under it), and leave 1″ between the top edge of the highest course and the ridgepole (so air can waft up into the ridge vent).

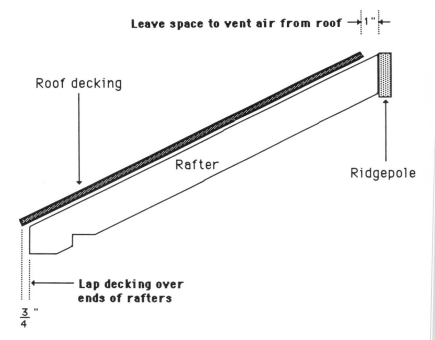

Roof decking

Drip Edge

Whether or not you put it on depends on how worried you are about the weather wheedling its way into the outer edges of your plywood roof deck. It's cheap, though, and relatively easy to install. It comes extruded in many ways, but you only need the simplest kind—a strip of aluminum bent to an L-shaped cross section. If you have any leftover 8″ flashing and decent tin snips you can even make it yourself. Cut the flashing into three $2\frac{2}{3}''$ strips, then fold each strip over a board so that one leg is about $\frac{3}{4}''$ long. The short leg of the L laps over the edge of the plywood, the long leg is stapled to its face. Do only the side and bottom edges of the roof. The ridge edge will be covered by roofing and your ridge vent.

Felt

Use the 15-pound weight. Start by stapling the bottom course so it laps about 1″ over the edges of the roof, and don't overdo the staples. Try getting a couple of thick beads of roofing caulk under each overlap, especially the first one. (Like many roofing materials, roofing caulk is black, tarry, and altogether repugnant to work with—unless you like that sort of thing. It's packaged in standard caulking tubes, though—Black Jack is a well-known brand—which makes it a shade less disgusting than the slop that comes in cans and has to be troweled on by untouchables.) Lap each of the subsequent courses about 6″ over the one before. Cut the top course so it doesn't block the vent opening, or, better yet, lap it however much it needs to be lapped so you don't have to cut it.

Roofing

You can use roll roofing, which is what you'll see in subsequent illustrations, or asphalt shingles. Cedar shingles prefer to be nailed to furring strips (horizontal strips of wood with a few inches of breathing space between each strip) instead of plywood decking, and they also like to be interleaved with 30-pound felt instead of the usual 15-pound. If you're wedded to them, there are dozens of books that adequately explain how to install a wood-shingle roof, which, at this stage of your odyssey through the mysteries of construction carpentry, you're certainly competent enough to do. As for asphalt shingles, nowadays they vary so much from brand to brand that it would be hard to take you through all the details of their installation. They're sold by the "square" (enough to cover 100 square feet of roof), and nearly all of them are packaged with good, clear instructions. They come in many grades and colors, and some even have a strip of adhesive built into their undersides to help hold them down. (It's still mainly roofing nails that hold them down.) They're all made of asphalt-impregnated felt, but in the better ones the felt is partly composed of fiberglass, which deteriorates more slowly than felt of the common sort. They're better-looking and more expensive than roll roofing and they don't go on quite as fast. The outbuilding seen in these pages is surrounded by woods, so it seldom has to dress for dinner. I therefore opted for roll roofing, which is also made of asphalt-impregnated felt. The kind I used is called "split-sheet" roofing because its surface is coated with mineral particles only for half of its 36" width. The coated half is exposed to weather; the uncoated half, called the selvedge, laps under the course above.

Installing Roll Roofing

To lay roll roofing, get a few pounds of 1" galvanized roofing nails (the ones with broad flat heads and rough-surfaced shanks). Cut a strip of 15-pound felt about a foot longer than the roof and fold it carefully in half the long way. The doubled roofing felt will be your starter course. Nail it along the entire lower edge of one side of the roof, and lap it about 1" past the plywood. Two rows of nails, the rows about a foot apart and the nails about 6" apart within the rows, should hold it down. (The purpose of the starter course is to compensate for the absence of selvedge from the previous course, since, in addition to the layer of felt you've already put down, you want either two thicknesses of roofing or one thickness of roofing and a double thickness of felt over every point on the roof.) When the starter course is down, nail a course of split-sheet roofing over it, taking care to line up its lower, mineralized edge with the lower edge of the starter course. Don't put any nails through the mineralized part, though. You want to avoid having any exposed nailheads, so nail only through the selvedge. And, as with the starter course and the courses that follow, make it lap at least 3" over the side of the roof. You can trim the sides more neatly after you've nailed on the gable fascia.

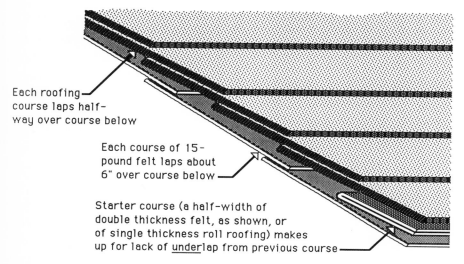

Each roofing course laps halfway over course below

Each course of 15-pound felt laps about 6" over course below

Starter course (a half-width of double thickness felt, as shown, or of single thickness roll roofing) makes up for lack of <u>under</u>lap from previous course

Roll roofing courses

Now, course by course, work your way up to the ridge. You'll find it's almost impossible to lay the roofing perfectly flat, but don't concern yourself too much with minor bumps. Over time, as the roofing softens in the heat of the sun, most of the bumps will go away all by themselves. When you get to the ridge, keep going until the mineral surface comes up to the ridge edge of the plywood. Use only one row of nails, and put the nails in fairly close to the plywood edge, so they'll eventually be covered by the ridge vent. You'll have to cut the selvedge off this last strip, but it won't be wasted. When you do the other side of the roof, you can use it as the starter course.

Incidentally, if the last length on the roll isn't enough for a full course, don't throw the partial course away. Just start another sheet right there, lapping the new sheet about 1' over the one that ran out and sealing the lapped surfaces with roof sealer. Speaking of which, after you're all done you'll notice that the mineralized half of each course isn't nailed down. Wind and water can therefore insinuate themselves between the sheets. So get your caulking gun and some tubes of roof sealer, lift up the edge of every course, and put a fat bead of sealer under it about 2" or 3" from the edge. Ugh!

Ridge Vent

Ridge vents are actually very simple affairs. They let fresh air enter the roof through vents in the soffits, rise up along the rafters, and exit at the ridge. (You vent a roof to release moist air that might otherwise cause condensation, which can eventually rot out the rafters.) There are any number of other roof ventilation systems—some are just screened-over holes, others call for continuous-duty exhaust fans—but modern ridge vents offer some distinct advantages: they're cheap to buy, easy to install, and, however and whenever you decide to finish the interior of the outbuilding, they impose a minimum of limitations. They're usually 8′ strips of light-gauge sheet aluminum extruded so they look like long skinny awnings with louvered undersides. The bottom skirts have little drain openings in them and are pre-punched for nails. The ends of the sections are mated to fit one into the next, which allows you to make a continuous vent of any length, and the sections can easily be cut to size with tin snips. You close up the ends with a pair of matching end caps that pop easily into place, and you nail them down over

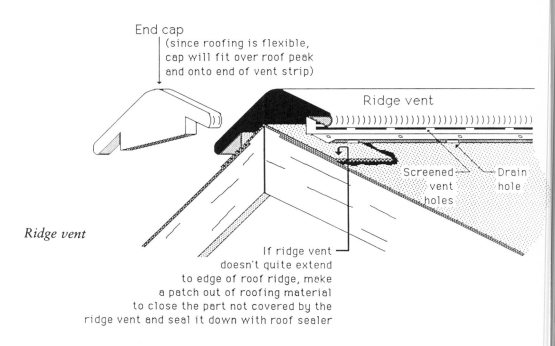

Ridge vent

the ridge with annular nails (they have threadlike grooves in their shanks to keep them from working their way back out of the wood) that are usually packaged with the ridge vent. Your lumberyard can supply you with what you need. When you install the vent—not a demanding task—you might want to put a bead of roof sealer under it before you put in the nails, and you also probably want to put a little dab of sealer over each nailhead. Then squeeze some additional sealer over any seams that might conceivably ship water. Your roof will now be weathertight.

Soffits

Along each of the long sides of the outbuilding, nail a soffit nailer to the siding so its underside is level with the soffit cuts of the rafters. Use whatever scrap lumber you have around. Anything with a $\frac{3}{4}$ x $\frac{3}{4}$ cross section or larger should do fine. The nailer will disappear into the innards of the roof. Its only purpose is to give you some nail bearing (i.e., a surface to nail to) for the edge of the soffits that butts the siding.

Then make the soffits. If your overhang is small enough, you can make them out of whatever you're using for exterior trim, in which case you should decide what kind of boards you'll trim the outside of the outbuilding with. This means soffits, fascia, corner boards, and a few small odds and ends. In practical terms the decision is basically between pine and cedar. Pine is cheaper, but it really ought to be painted. Cedar, which comes either smoothly milled or intentionally rough-sawn, contains enough natural oils and resins to stand up to the elements as is. Both are easy to work with, and the main reason I chose cedar myself is that it doesn't need maintenance. Whichever you decide on, the most commonly available boards are "one-bys" (1x6, 1x8, etc.), which means they're $\frac{3}{4}''$ thick.

If you can't get boards wide enough to span your overhang, make the soffits out of plywood ($\frac{1}{4}''$ is thick enough). If you have any leftover T-111, you can even use that; just cut the strips so they span the horizontal distance between the outer corners of the rafters and the siding. If you're out of T-111, you can probably get plywood to match the rest of your exterior trim (that is, plywood whose outer ply is paintably smooth-milled or made of rough-sawn cedar).

Before you nail the soffits on, though, you have to vent them so air can flow up through the roof and out the ridge vent. Use circular aluminum louver vents at least 2″ in diameter, preferably bigger (they're available in $\frac{1}{2}''$, sometimes even $\frac{1}{4}''$, gradations). Find a circle cutter for your drill (throughout this book, the word "find" in contexts like these is a glyph for "borrow if you can, buy only if you have to") and drill out the holes for the vent inserts (which come in cardboard boxes that tell you how big the circle should be). Try to keep the holes evenly spaced and their centers in a straight line. One hole between each pair of rafters is enough, and they don't have to fall midway between the rafter ends. All they need to do is admit air into the bay. And don't put in the vent inserts until after the soffits are up. If you do it now they'll probably pop out.

Now nail up the soffits with 7d or 8d galvanized finishing nails (the ones with very small heads). In doing so, you'll probably discover that the outer corners of your rafters don't describe a very straight line and the outer surface of your siding isn't all that straight either. Don't be alarmed. There's more trim coming.

Soffit with vent holes

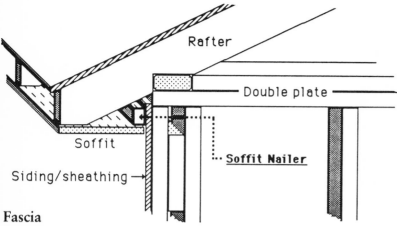

Fascia

Butt it tight up to the roof decking (you may have to work it under the drip edge), then nail it along the fascia cuts of the rafters. Galvanized 8d common nails are best. Make the fascia from whatever exterior trim you've decided to use. If you rip it down from wider boards, keep the machine-milled edge facing downward. That way your own humanely irregular rip cut will be out of sight. Rip (or buy) the fascia wide enough so that it extends 1" or so below the underside of the soffits. If there have to be any butt joints along the length of the fascia, try setting your saw to make 45° bevel cuts.

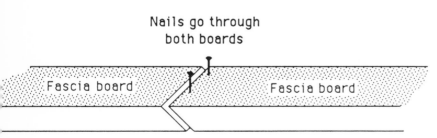

Matched 45° bevel cuts

Bevel butts look nicer over time than right-angle butts. And try arranging it so that any joints fall directly over a rafter end, so the nails that go through the fascia boards continue on into the rafter.

Now do the fascia boards for the gable ends. Any width is good as long as the fascia covers whatever irregularities there may be in the upper edges of the siding. Beyond that, aesthetics supersede utility. Fabricate the fascia from whatever you're using for exterior trim, and measure carefully enough to make the lower ends of the boards line up well with the fascia that's already on. And jam the boards as tight under the overlapping roofing as you can get them. Nail them with galvanized common nails of a size commensurate with the dictates of your developing intuition.

Corner Boards

Once the fascia's up, cut some narrow strips of felt, staple them over the four exterior corners of the outbuilding, and then nail on the corner boards. The ones on the "long walls" extend from just under the soffits to the lower edge of the siding. The tops of the ones on the "short" walls are cut at an angle so that they butt the gable-end fascia cleanly. Again, make them from whatever exterior trim you've chosen, and, using galvanized nails, nail them so that they hide the felt you've just tacked onto the corners.

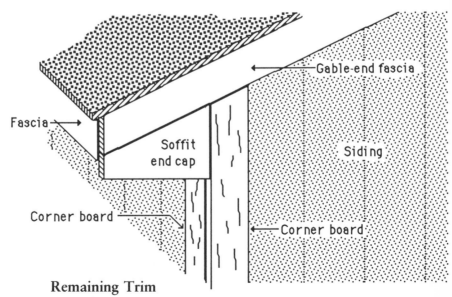

Corner boards and soffit end cap

Remaining Trim

With scraps, make some end caps to close up the soffit ends. They're highly visible, so a certain amount of care is warranted. Nail them with galvanized finishing nails.

Finally, if you've done any ripping of exterior trim, there's a good chance you've inadvertently made some molding. If you haven't, buy whatever kind you like. Then nail it up under the soffits to cover the seam between the soffits and the siding, and, if you like, nail some short pieces under the insides of the end caps. Use galvanized finishing nails once again.

Now put in your circle vents. They're made for friction fit, but if the holes you've drilled for them are too large you can "glue" them in with discreetly small amounts of caulking.

Your roof should now be sufficiently well sealed so that anything capable of getting in would be too small to do it—or you—any harm. It will stay dry and still be able to breathe. The nail ends of your roofing nails are visible from inside, but that's the way the underside of a roof is supposed to look. At last your outbuilding genuinely deserves to be called a building.

10

Ingress: Decks and Stairs

You're probably tired by now of turning your ankles on the concrete block or temporary plank that you've been using to hoist yourself up to the door, and making exterior stairs and at least a rudimentary landing will give your outbuilding a welcome civilizing touch. If you're feeling ambitious, the landing can be a full-blown deck or even a covered porch. If you're not, all it has to be is an enlarged uppermost step. Whatever you decide on, making simple stairs and decks is a good way to use and appreciate the proficiencies you've been acquiring.

DECKS

All a deck really is is a boxed matrix of rafters covered over with deck boards. The underside of a small deck might look like the illustration at right, above.

One edge of the deck is nailed to the building so it butts tightly up to the underside of the door threshold (right, below) while the rest reaches out into the surround. The joists can run either parallel or perpendicular to the building, normally whichever way keeps them shortest. The deck boards are nailed across the joists, the joists are supported at their ends by the boxing, the boxing is supported on one side by the building and on the opposite side by posts, and the whole thing comes together very fast.

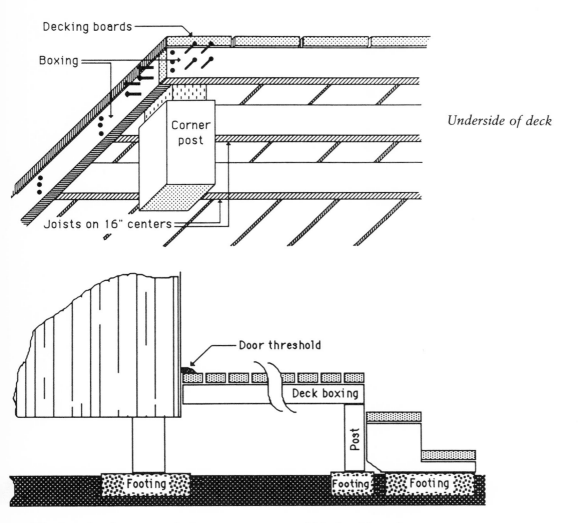

Underside of deck

The best lumber to use is the pressure-treated kind, with cedar running a respectable second. (Redwood is great, but it's expensive.) The joists and boxing can be anything from 2x4s to two-by-whatevers, with the understanding that the heavier the lumber is, the bigger your deck can be without having to run girders under the joists. If you use 2x4s, no span along the joists or boxing should exceed 5'. For 2x6s, the maximum span is 8', and for 2x8s it's 11'.

Usually 4x4s are the stuff of deck posts, but if you've got any leftover lengths of pressure-treated 6x6 and don't mind their somewhat stout appearance you can use those too. As for cutting the posts to rest the boxing on, I suggest you do them like this:

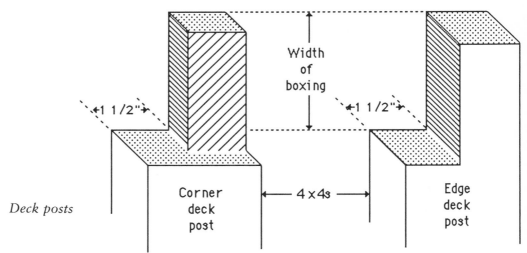

Deck posts

The deck posts should all sit on footings, but deck-post footings needn't be anywhere nearly as elaborate as the footings for the outbuilding itself. If you like playing with concrete (not so aberrant a fondness; there's satisfaction to be had in making stone out of slop), you can dig holes about 1' wide and at least 4" deep and, after mixing concrete in the way we've already discussed, simply fill the holes to a level slightly higher than the grade. If that approach seems too amorphous to you, you can make footing forms out of scraps of 2x4 and put them in the holes before you pour. After the concrete hardens for a day or two, just knock off the forms and backfill. But the easiest footings of all are concrete blocks sunk in sand (or sandy soil) so their top surfaces are a little above grade. They're cheap and require minimal effort to put in, and even if they sink a tad with time, the worst you'll have to do is shim the posts from underneath in a few years. Regardless of what kind of footings you make, cut a piece of flashing to put under the bottom of each post. It'll prevent moisture from wicking out of the footing and up into the wood.

As for deck boards, I'm an advocate of 2-bys (2x4s or 2x6s). Some people use thinner, but I think the solidity underfoot and additional durability are worth the extra money. The boards

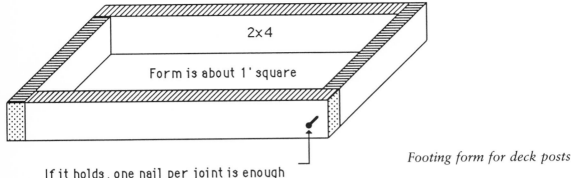

If it holds, one nail per joint is enough

Footing form for deck posts

mustn't be butted tightly up against each other when they're nailed, since rain has to drain between them. So use loose 10d, 12d, or 16d nails as spacers. The thickness of a nail is all the drainage room they need.

And if your deck is long enough so that the joists run on for any appreciable length, cut pieces of solid bridging to nail between them. Apart from safety considerations, they'll feel better to walk on.

Still another "if": if you want your deck to have a neat, polished appearance, set your saw to cut at a 45° angle and bevel-butt the corners of the boxing.

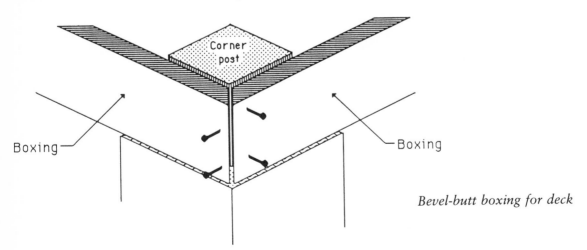

Bevel-butt boxing for deck

There really isn't much more you need to know to make a deck. I'd advise that you make the boxing first, then attach it to the outbuilding and temporarily support the outside corners. If the deck is just a landing no wider than the threshold of the door, simply nail the deck boxing to the outbuilding boxing under the door so that there's just enough room to wedge a deck board between the underside of the threshold and the upper surface of the boxing. If, on the other hand, the deck runs along the building for a while, you can attach it in one of three ways.

Attaching the Deck to the Outbuilding

Way one, the simplest, is to nail the deck boxing right through the siding and into the building boxing, again leaving just enough room for a deck board to fit tightly up under the door threshold. (This far along into the project I need hardly remind you to level the deck boxing before you nail it.) Yes, there'd be those who'd raise objections to way one on the grounds that so much exterior, unventilated wood-to-wood contact might eventually rot away the siding just behind the deck boxing. How right or wrong they'd be is pretty much a judgment call, but, to placate them, you could adopt way two.

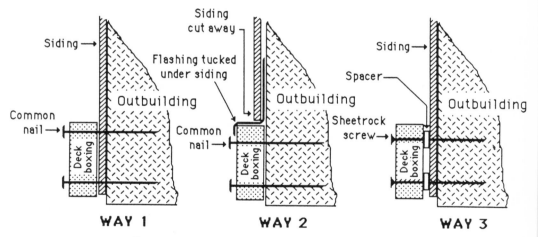

In way two you set your saw to cut at a depth equal to the thickness of the siding, then cut away the siding where the deck butts the outbuilding (making *very, very* certain that you don't cut into any nails), work a strip of flashing in under the bottom of the siding where it's been cut away, nail the deck boxing to the *building* boxing, lap the flashing over the deck boxing and finally renail the siding just above where the deck meets the outbuilding.

If that sounds like a lot of work, it is, especially if your deck's a big one. Which is why I recommend way three. In way three you use screws—long Sheetrock-type screws. Except you fit a spacer onto each screw so that the deck boxing is held about $\frac{3}{8}''$ away from the siding. You can make the spacers by stacking metal washers to $\frac{3}{8}''$ thickness or you can use ordinary machine nuts big enough to slide over the screws, perhaps sandwiching each nut between a pair of washers first. (It's best to use rust-proof hardware—brass, stainless, or plated.) Now there'll be room for air to move around between the deck boxing and siding, the boxing will be quite secure, and you won't have to fuss with flashing, reset the depth of your saw cut, or any of those other noisome things.

That done, construction of the remainder of your deck ought to be pretty much self-evident.

STAIRS

When you've got a deck—even if it's just a mini-landing—you'll need some stairs to link it to the ground. Simple ones. People don't expect exterior stairs for outbuildings to be as intricately carpentered as interior stairs for houses. All you'll have to make are treads (the actual steps), stringers of the cutout sort (the members that hold up the treads), and perhaps a banister. In many areas, building codes require that any staircase with three or more steps have one, but a 2x4 pitched to match the pitch of the stairs and set at a comfortable height above them (say 2′6″) is usually all you need. I'll leave the design of the banister to you, but remember to support it with posts no more than four steps apart. The posts can be 2x4s too, and you can nail their bottoms directly to the staircase.

Some Nomenclature

Like roofs, stairways have rises and runs. The total run is the horizontal distance between the outer edge of the stairway and the edge of the deck. The total rise is the vertical distance between the top of the deck and the point on the ground where the outside of the stairway comes to rest. The *unit* rise is the vertical distance between the top of any tread and the top of the next tread; the unit run is the horizontal distance between the outer edges of those same two treads.

Materials

Since exterior stairs with a unit run of much less than 9″ don't feel very comfortable, I'd suggest you make the treads out of sections of 2x10 or, to go grand luxe, 2x12. You can also use a pair of 2x6s for each tread, or even three 2x4s. If you do that, remember to keep a nail's worth of space between the boards when you fasten them to the stringers.

The stringers are also best made of 2x10 or 2x12; I personally prefer the latter. Stringers can butt into the deck in one of two ways:

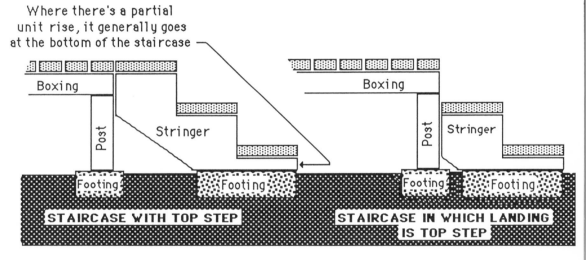

Stringers

I know of no special advantage that one way has over the other; your choice will depend partly on looks, partly on how many 2x10s and/or 2x12s you feel like buying and cutting, and partly on the fact that the stringers for a staircase *with* a top step can be toenailed directly to the deck boxing (and therefore don't require that there be posts immediately nearby), while stringers without a top step have to be toenailed into posts (which may necessitate your installing posts where you'd otherwise not want or need them).

Designing a Staircase

To determine the number of stairs you'll need, start by deciding on a unit rise. A practical maximum is 8", 7½" feels more or less normal, and 7" or less is even better. Then divide the unit rise into the total rise. The result will be the number of stairs you need, but it's not quite as simple as all that.

Say your total rise is 32" and you'd like ordinary 7½" stairs. If you divide 32 by 7.5 you get 4.267, which means you'd have four full 7½"-rise stairs and a fifth stair a little over one-quarter height. But .267 × 7½ is just about 2", and, if your tread, being a 2-by, is 1½" thick, that would mean that the part of the stringer *under* the tread would be only ½" thick, which virtually guarantees that it will eventually break.

There probably ought to be a minimum of 2" of stringer under the bottom tread, so what's needed is some engineering (i.e., playing around until things come out more to your liking). If we change the unit rise to 7" and now divide 7 into 32, we get 4.57. Now you'll still have four and a fraction stairs, but the bottom step will be .57 × 7" or just about 4" high. Of those 4", 1½" is accounted for by tread thickness, but there'll now be 2½" of stringer under the tread, which will be much less likely to snap off than the ½" sliver you arrived at first time around.

Cutting the Stringers

To cut the stringers, use your carpenter's square to make the marks.

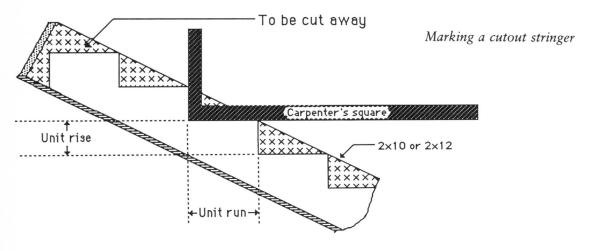

Marking a cutout stringer

Remember to allow for the thickness of the treads when you mark the tread cuts at the top. In fact, because cutting stringers often brings out the part of you that's particularly error-prone, you might want to make template stringers out of plywood or even cardboard first, then test them with real treads to make sure all the rises come out right before transferring the marks to your archival stringers. If you like long treads (4' or more across), make a third stringer to run between the outer two. Otherwise the stairs will be objectionably bouncy. Use galvanized nails for all exterior carpentry, and put a strip of flashing and a concrete footing under each stringer's lower edge. (A brief note on flashing: copper is a popular, if expensive, alternative to aluminum, but don't use it if you're flashing cedar or redwood. Copper finds the resins in these woods chemically disagreeable and responds by corroding rapidly.)

AN OPTIONAL PROJECT

As yet another gesture to civilization, or maybe just an exercise, you might want to build an entry awning for your door. Its frame could look something like this:

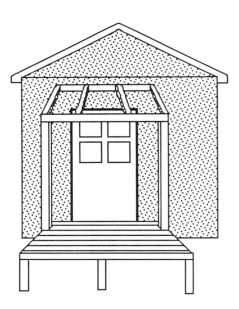

It's shown butting into a "short" wall, but it can butt into a "long" wall just as well. It needn't be pitched as steeply as your roof, and if it *is* built along a long wall, it can butt into the siding, the fascia, or the roof; its height and pitch will govern which of those it is. It can be topped off with the same combination of decking, felt, and roofing that you've installed on the roof, or it can simply be covered with inexpensive corrugated fiberglass, which you can buy at hardware stores, lumberyards, and other building-supply sources.

If it butts the siding or the roof, it will be happiest if there's some flashing along the seam, so that water is conducted down onto the awning and out of harm's way. Earlier in this chapter, when we discussed some different methods for attaching your deck to the outbuilding, I described a way to install flashing under the siding and called it "way two." Consider way two if the awning butts into the siding, and make sure the flashing drains onto the top of the awning's covering. If the awning butts the roof, things are much simpler. All you have to do is run some flashing from under the next highest course of roofing down to the top of the awning.

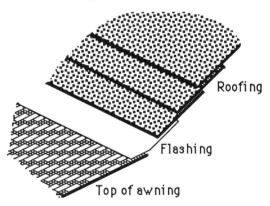

If the awning butts the fascia of the outbuilding, you can skip flashing altogether. The roof will drain down onto the awning without any further help from you.

You can frame the awning all in 2x4s, with a possible concession to 4x4s when you're making the posts. And if you're using real roofing to cover the awning and you want it to look especially spiffy, get some clear plywood and cut a soffit for underneath the awning rafters.

With those things in mind, you should be able to do the rest of the design and construction on your own. If I were to tell you any more, the project wouldn't be an exercise.

Now, with your stairs and landing in place, not only do you have a totally enclosed outbuilding, you can even get into and out of it in the way one customarily enters and exits the most genteel of dwellings. At least you can by day, when you can see. Yes, by night there's always the extension cord that you've been using for your power tools, but darkness seems to fall with special weight on buildings that don't have a single working electrical outlet.

11

When to Stop

Of the several milestones we've arrived at, the one we're perched on now is a good place to pause and ask when enough is enough. You already have a structure capable of keeping out wind, rain, and bugs, of admitting light through its windows and people through its door, also of excluding light and people if that's what you want, and of standing proud and tall for many years to come. You'd probably like a more convincing floor than the (by now) battered plywood platform you've been walking on, and there are still those pesky electrical outlets to be taken care of. (We'll get to both in due course.) But by and large the outbuilding is a pretty decent piece of shelter as it is.

Oh, you could grouse about the lack of interior walls or window trim or baseboards or a handful of other items from the traditional inventory of domestic amenities, but you'll have to grant that one thing you don't lack—please try not to gag at the phrase—is rustic charm. Because somewhere at the aesthetic center of the whole notion of rustic charm is the fact that at any given moment, you can look around the shelter you're in and see how it's going about the business of sheltering you. In a seamless, highly finished interior, everything's concealed. The stuff inside the walls has been relegated to the realm of mystery. All the skeletons accumulated during the building's construction have been shoved into a closet, and the closet's been nailed shut and walled over.

By contrast, what gives the rustic style so much of its appeal is the avoidance of concealment. In rustic houses virtually all the innards are laid bare. When you want to make a finished room look rustic you expose its beams and strip its walls down to brick. Far from holding any mystery, rustic interior walls are like a blueprint of the house. You can look around and see how it was built, often to the point—if you look hard enough—of getting a feeling for the personality of whomever it was who built it. Seamless, highly finished interiors can be chillingly anonymous; rustic ones, even when they're relatively new, seem to have a signature, a personal imprint, a redolence however slight, of history. They put their imperfections boldly on display, and the result can be affectingly human. Rusticity can also, when it's overdone, be excessively sentimental, but as a guiding approach to the look of an owner-built outbuilding it isn't to be sneezed at.

Unless, that is, you're sitting in it in the dead of winter, sneezing like hell, trying to accomplish something, and wondering why your heat source is groaning impotently under what seems to be so minimal a load. Your bare-bones framing may be crying out to please you all over again by doing service as decor, but down around zero degrees a cozy layer of insulation nestled behind smooth Sheetrock walls can be warmer than all the rustic charm in Christendom.

So yet again you're faced with a decision—whether or not to insulate and wall up the inside of the building, whether or not to surrender the rusticity that's already yours in hope of getting back an equal measure of year-round practicality, whether or not to do a whole bunch of additional work when your hands, your schedule, and your checkbook are all singing *basta!* in three-part harmony.

My advice is this: if *basta!* is indeed the dominant refrain, call it quits for now. Insulating, Sheetrocking (or paneling), and trimming the building can be postponed until you're ready for another project. Do put in a floor, if only to protect the subfloor. All it needs to be is a neat layer of plywood covered with a coat of paint. When you install electrical outlets you can use one of several forms of surface wiring, all of them eminently serviceable. (See Chapter 12 for some options and considerations.) And take a few moments on a very bright day to blind the windows, plant yourself inside, and look for light leaks. If there are any tiny holes or discontinuities in the siding, they'll leak not only light but air and maybe bugs and water too. Patch them with caulking compound and, without going any further, your outbuilding will probably be tighter than most of the known world's extant dwellings. It will be fit for year-round use in warmer climates (where the only heat that's needed is a desultory blast or two from an electric, kerosene, gas, or wood heater) and for summer, early-fall, and late-spring use in colder climates (where, once again, the only heat that's needed is the same occasional blast from the same electric, kerosene, gas, or wood heater.)

Actually, one very likely function for the outbuilding that hasn't received its fair share of attention is the whole question of storage. Because if what you need is primarily a mini-warehouse, look no further and *build* no further. You already have what you're looking for. (And if it *is* a storage shed you've built, I hope you've had sense enough not go overboard on expensive doors and fancy windows; the chances are excellent that your belongings will be just as happy in the dark.)

Even if you expect to spend somewhat more time in the outbuilding than that, I strongly advise taking a good hard look at the question of insulation. Insulation inhibits heat loss through radiation, but it's not especially effective against the infiltration of cold air, which remains the chief means through which most buildings give away their heat. What with the virtues of insulation having been trumpeted so loudly in recent years, many people have come to believe that uninsulated buildings are flatly unheatable, and it simply isn't so. The outbuilding you've been seeing pictures of *is* insulated; it's used throughout the winter, and it's heated by a smallish wood stove whose efficiency is less than stellar. Yet, even when the weather's cold, heating the building comfortably isn't anywhere nearly as much of a problem as getting the stove's air regulator closed down enough to keep the occupants from roasting.

Given these and other factors (factors peculiar to your own anticipated usage patterns that I daren't even guess at), the decision about when to stop clearly bears some thought. You might, for instance, want to insulate the roof and not the walls. It's through the roof that any building loses the bulk of its radiant heat, and the insulation will also cover all the nail points poking down through the roof decking, which may serve to quiet speculation that your taste in architecture runs to School of Torquemada. You can then Sheetrock or panel over the ceiling and still let your framing and the inside surface of the siding make do for interior walls, thereby preserving the rough-hewn look of your interior.

Or you may want to Sheetrock or panel both your ceiling and your walls, not insulate either of them, and wait to see how things go. Installing fiberglass insulation isn't quite the breeze it appears to be, and you may not want to take the time or spend the money. With interior walls you'll be able to use ordinary inside-the-wall wiring techniques when you power up the building (see Chapter 12), and later, if it turns out to be hard to heat or the way in which you use it changes, you can always bring in one of those people who blow loose fill insulation into the walls. There remains the chance, especially in milder climates, that the interior walls will be all the additional insulation you need, since the "dead air" between the inside and outside walls is itself a form of insulation. On the other hand, if your summers are very hot and you expect to run an air conditioner a good deal, insulation will keep *out* the heat and accordingly cut down on your electric bills.

Which is all by way of saying that you have a multiplicity of options, and that you'd do best to regard the following chapters as a kind of menu from which you can select according to the dictates of your appetite for construction and your ambitions for the building. If you're going to use it heavily in winter, by all means insulate the dickens out of it. If you've already begun seeking treatment for builder burnout, take every shortcut you possibly can. If you're a glutton for punishment, do everything the book suggests and then maybe get a router and a set of bits and take a few extra months to filigree the trim both inside and out. All you can do is screw up, and since it's fairly well established that to err is human, your errors may infuse the building with a certain rustic humanity. In fact, if you're excessively perfectionistic and loathe the very thought of error, maybe you ought to set yourself the task of building a conspicuously imperfect outbuilding just as therapy. The possibilities are endless.

12

Power

In early childhood many of us were taught not to be hoodwinked by the apparent usefulness of electrical outlets. However benign they might have seemed as they squinted at us from their warrens in the walls, at bottom they were agents of catastrophe, eager to zap anyone who came too close. They liked especially to go for fingers, wet ones most of all. But it made no difference where they got you. Once the current flowed you were a goner, the victim of a sudden and horrible death made even more tragic by the fact that the killer had been living in your very midst.

That's what I was taught, and yet I've always enjoyed puttering with electricity. I've never found it very fearsome, but a few chapters back I mentioned counterphobia, and it's possible that anything my mother claimed was dangerous took on particular appeal. One of the more intriguing perils she always liked to emphasize was that if you touched anything "hot" you wouldn't be able to let go. And in the early days of electrification, when nearly all houses were wired with direct current, it was true. At household voltages, DC glues you to the point of contact, so you not only die but you die in chains. Which is why alternating current has become the national standard. (Actually, apart from batteries, battery chargers, and special generators, it's hard to find DC these days. If you need it, you usually have to convert AC with a rectifier.)

Nonetheless, I know some perfectly normal adults from whose

complexions the mere sight of an exposed wire will drain a shade or two of color, and if you're one of them, by far the most prudent thing to do would be to go out and hire an electrician. I can understand that taking so drastic a measure might seem to violate the very essence of this book, but it doesn't at all. Part and parcel of the book's ethos is that there's no special virtue in being insanely rigorous. Of course, professional electricians come at a steepish price, while friends familiar with the theory and practice of domestic wiring (nearly everybody's got one) can usually be had a lot more cheaply. Persuading such a friend to—how shall we put it?—share his expertise would be field-expedient in the noblest sense and is therefore altogether consonant with everything this book espouses.

On the other hand, if you already know your way around a circuit, or you're willing to find a good book or two and make your way (there being insufficient room here for a comprehensive cram course on house wiring), there's no reason why you can't do it all yourself. It's not that big or complicated a job, and it's a nice change of pace from carpentry. You'll also have the chance to put outlets where *you* want them, not where some anonymous contractor thought someone *might* want them. And if you're worried that your work will have to pass inspection, try viewing your inspector as an asset. The chances are he checks out owner-installed wiring all the time, and he can tell you where you've flubbed (if you have) and even point you back in the right direction. The carpentry in your outbuilding will probably hold together whether it's inspector-approved or not, but if you aren't confident about your electrical skills it's a good idea to have someone fairly knowledgeable—if not an inspector then a licensed electrician or a reliably savvy amateur—examine what you've done. This is your big chance to *create* the mysteries inside the walls instead of being baffled by them, so there's every reason to get some certification that your mysteries are sound ones.

WHAT'S INVOLVED IN WIRING

Look over the following list, then, with an eye toward what's within your scope and what's beyond it. Think about which things you'll do yourself and which you'll farm out. Very broadly, this is what needs to be done to power up the outbuilding:

1. Designing the circuit
2. Deriving a materials list from your design
3. Buying materials
4. Putting wall boxes where the interior switches, receptacles, and light fixtures go
5. Drilling holes in the siding (of about $\frac{5}{8}''$ or $\frac{3}{4}''$ diameter) where the *exterior* switches, receptacles, and light fixtures go
6. Putting wall boxes wherever there are additional circuit junctions (all splices have to be contained within a wall box)
7. Drilling the studs to create wire routes in the walls
8. Stringing wires between all the boxes and exterior holes
9. Installing an additional breaker in your house's breaker panel (or, if it's an older house, a new fuse circuit in its fuse box)
10. Running exterior-grade cable between your house's breaker panel (or fuse box) and the main junction box of the outbuilding
11. If you want to be able to switch an outside light from both the outbuilding and the main house, installing a three-way switch somewhere inside or outside the house and a second three-way switch somewhere inside or outside the outbuilding, then running wire from some power source in the house to the first three-way switch, on to the second three-way switch, and finally to an exterior light fixture on the outbuilding
12. Hooking up the loose wires dangling from the boxes (they're called tails) to the switches and receptacles that go *into* the boxes
13. Screwing the switches and receptacles you've just wired up into their boxes
14. Threading exterior-type boxes onto the exterior tails
15. Screwing the exterior boxes onto the outside of the building

16. Caulking the exterior tails around where they enter the exterior box
17. Hooking up the exterior tails to their fixtures (outlets, switches, light fixtures)
18. Screwing the fixtures into their exterior boxes and screwing the cover plates down onto the boxes
19. Connecting the wires in the junction boxes
20. Perhaps running exterior-grade telephone lines and/or 75-ohm shielded television cable between the house and the outbuilding
21. Perhaps installing phone jacks and/or cable TV jacks in the outbuilding and hooking them up to the phone and cable lines
22. Checking everything carefully
23. Testing what you've done
24. Trenching and burying your exterior cables

Now, if all of that doesn't daunt you (and there's really no reason why it should; it all goes very fast), decide where, inside and outside the building, you want your outlets, switches, and light fixtures, and then plunge into the project keeping the following considerations in mind.

Surface Wiring

Even if you're not going to have interior walls, there don't have to be any visible rats' nests of wires. Several surface-wiring systems are available (two well-known brands are Wiremold and Plugmold) in which all the fixtures—switches, junction, receptacles, light fixture boxes, even wire runs—are made to fit modularly together while giving a neat appearance and conforming to most electrical codes. If they're okay with your inspector—and they'll probably be fine—I'd suggest you go with them. They come with copious instructions, and you'll be able to ignore much (but not all) of the following.

Outlet Heights

Electrical codes stipulate that the bottoms of wall outlet boxes have to be at least 12" off the floor, and what most installers do is measure with their hammers. Hammers are typically 13″ long, and the boxes are then said to be at "hammer height." Don't put them higher than 18″ off the floor unless they'll service a counter, in which case 4′ is a good height; 4′ is also code height for wall switches.

Junction Box

Since there can't be any unboxed junctions in the walls, install a separate box (the 4″-square kind) at hammer height near where your service cable will enter the building and use it as a junction box. When you're all done building, you'll close it with a solid cover plate.

Outlet Spacing

Electrical codes usually require that receptacles be no more than 12′ apart along a wall, although, to my taste, that's being rather stingy. (Incidentally, when you measure along a wall, you don't have to include doors.) Do yourself a favor and space them more closely. Where you plan on having a work table or desk, allow for two receptacles in one box (thereby giving yourself four outlets instead of the basic two). Code also stipulates that there be a light switch on the wall near the entry side of any door that opens into a room. The switch must control either a light fixture or an outlet (the idea being that you'll plug a lamp into the switched outlet).

Outside Lights

Code usually mandates that there be some kind of outside light near every exterior door and that there be a switch near the door that controls the light from inside. Code or not, I think it's a good idea. The best solution is to use a pair of three-way switches, one in the outbuilding and one at the main dwelling. That way you can illuminate your nightly journeys to and from the outbuilding, and the light can draw power from a circuit in the house, which will reduce the load on the outbuilding circuit. Three-way switches are available in just about all hardware stores, and most are packaged with a wiring diagram. But if

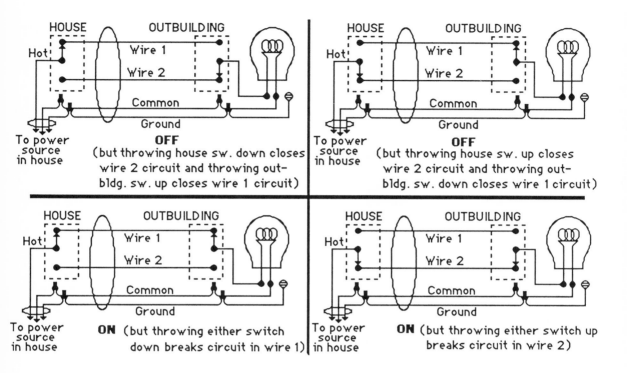

Schematic diagram of how three-way light switch circuit works

you don't know how the circuit works and you're curious, I've provided a schematic.

Schematic notation for switches is an arrow pivoting around an input contact, and, in the diagram, you can see how the switch selects one output contact when it's up and another when it's down. You run three-conductor (plus ground) wire between the switches, and if one switch is indeed at the house and the other at the outbuilding, the plastic sheathing should be of outdoor (i.e., waterproof) grade. The outer sheathing of waterproof wire is stamped "UF," and, curiously, it's often a bit cheaper than indoor-grade wire, which is stamped "NM." As for the gauge of the wire, it depends on the rating of the fuse or breaker that governs the circuit you're tapping for power. If it's a 20-amp fuse or breaker, use (at least) #12. For a 15-amp circuit you may be able to use #16, but if you want to run #16 UF cable outdoors, check with your inspector first. This is an area in which electrical code is subject to a degree of interpretation.

Wire Gauges and Color Codes

Actually, in the event that you're comfortable with house wiring but maybe a little rusty, let's review standard wire colors and gauges. Ordinary in-the-wall wiring is generally done with Romex-type cable.

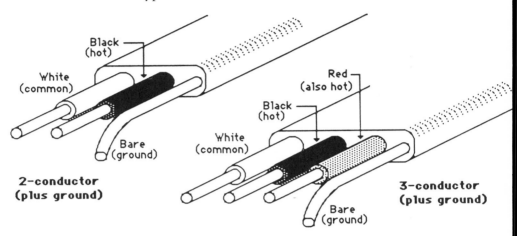

Inside-the-wall Romex-type cable

The two kinds you'll be likely to use are the two-conductor (plus ground) for the majority of your circuits and three-conductor (plus ground) for three-way light switch circuits. In two-conductor cable, the hot wire has black insulation, the common wire has white insulation, and the ground wire has no insulation. In three-conductor cable there are two hot wires, so one is black and the other red. The two gauges you're likely to use are #12 and #16 (12-gauge is thicker than 16). As for which to use, remember it's got nothing to do with a circuit's projected load. Wire gauge is entirely a function of the size of the breaker or fuse that governs the circuit. If the circuit's on a 20-amp breaker, use 12-gauge wire; if it's on a 15-amp breaker you can use #16 or larger. Twelve-gauge two-conductor cable has "12/2 WITH GROUND" stamped into the outer plastic sheathing. Similarly, 16-gauge three-conductor cable is stamped "16/3 WITH GROUND." The same goes for "12/3 WITH GROUND" and "16/2 WITH GROUND."

While we're on the subject of color codes, the screw terminals on receptacles, switches, and many (but not all) light fixtures are also color-coded. Brass screws connect to hot wires, chrome screws connect to common wires, and green screws connect to ground wires. Consequently, when you hook wires up to fixtures, you connect black wires to brass screws (the mnemonic is "b" goes to "b," where "b" stands for *burning hot*), red wires also to brass screws (*red hot* goes to *burning hot*), white wires to chrome screws (*white/ice-cold/common* goes to *chrome/cold/common*), and bare wires to green screws (*ground* goes to *green*). Not all receptacles and switches come with ground screws, but try to buy ones that do. Many inspectors insist on them.

Boxes

Code and sheer good practice require that all connections, whether wire-to-fixture (via screw terminal) or wire-to-wire (via twist-on wire nut), be contained within an approved box. Modular surface wiring has its own containment system, but if you're using ordinary plastic-sheathed wire, you have to provide a box for every fixture. By far the easiest boxes to work with are the plastic ones that come prefitted with a nail to bang into the side of a stud and with knock-out holes to let the wires pass through. Boxes are available in several sizes, of which by far the two most common are the one-unit size (with room for a single two-plug receptacle or a single switch) and the two-unit size (with room for two switches or two receptacles or one switch and one receptacle). They also come in different depths, but since they're cheap, get them all at least $2\frac{1}{8}''$ or even $2\frac{3}{4}''$ deep. The deep ones are much easier to work with, and code lets you make more connections in them. (More on that in a moment.) To find out if it's okay to use plastic boxes in your area, you can check with your inspector or simply unscrew the faceplate from an outlet or switch box in a newish house nearby (it's virtually inconceivable that any electrical contractor who's permitted to use plastic boxes would use anything else). Generally, however, it's only in a few large cities that plastic boxes don't pass muster, and most people in a position to build an outbuilding at all are also in a position to use plastic boxes.

An exception to the above: use metal boxes for light fixtures. Lights generate heat, and metal is much less vulnerable to thermal deterioration than plastic.

Mounting Boxes on Studs

If you're going to have interior walls, you might want to scan the chapter on that subject and decide how thick the walls will be. When you nail boxes to the sides of studs, they should stick out past the stud edges for a distance roughly equal to the thickness of the walls. Later, when you cut holes in the walls for the boxes, the front edges of the boxes will be more or less flush with the walls.

BASIC CIRCUITS

Here's a schematic diagram of a good basic circuit for the outbuilding:

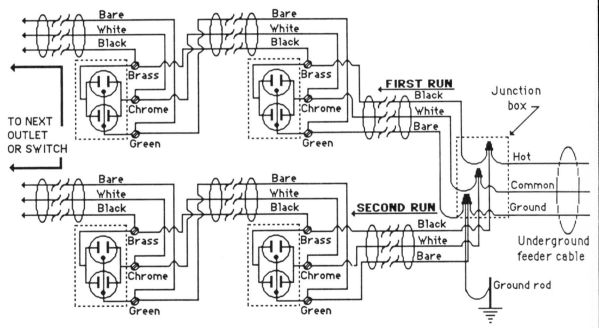

Schematic diagram of outlet wiring

You'll notice that there are two "runs" in the picture, although there's nothing wrong with doing everything in one run and simply wiring the feeder cable to an outlet, then wiring the outlet to the next outlet, then to the next and the next until you've wired up all the outlets.

You can tap any of the outlet boxes for a light circuit:

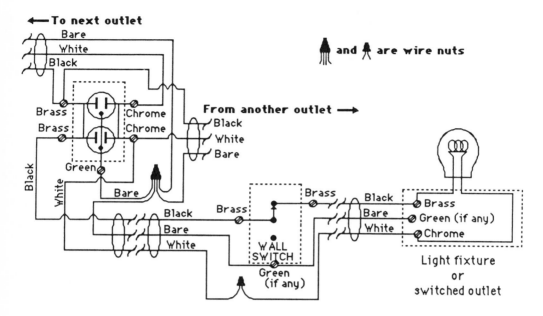

Common switched light circuit

The circuit's the same whether you're putting the light inside the outbuilding or outside: in one case you'll string cable to an interior box, in the other you'll fish it out through a hole in the siding. (That's where the value of schematic diagrams lies: being generic, they make fewer assumptions about your specific choices.) But a word of caution: don't tap any given outlet box for *more* than one circuit. Electrical code has a welter of rules governing the number of allowable connections in a single box. I won't run them down now because they're too complex, but a good rule of thumb is to have no more than three cables entering or leaving a box: an "in" cable, an "out" cable, and, if you need one, a "tap" cable. And don't twist more than two wires around a single screw terminal. Most receptacles give you two brass and two chrome screw terminals. They're yoked together with a strip of metal (brass to brass and chrome to chrome), so you have the luxury of an extra pair of terminals when you're wiring the receptacle. (Take a close look at the outlet fixture in the above schematic.) If you unyoke them by breaking off the metal strips with, say, a pair of longnose pliers, each of the two outlets in the receptacle will have its own pair of screw terminals, one brass screw and one chrome. You'll want to unyoke the pairs if you're connecting your light switch to an outlet rather than a light fixture. With the metal strips broken, each of the two outlets in a receptacle becomes a separate unit, so one can be switched while the other is permanently hot.

Ground Wires

Looking at the above circuit, you might have noticed that some of the ground wires were spliced together with a wire nut. That's because even though you get a pair of brass screws and another pair of chrome screws, you generally get just one green ground screw, and most inspectors don't like to see a tangle of ground wires all hooked to the same screw. The best solution is to twist-splice *all* the ground wires together with an additional short length of bare wire, bind the connection with a wire nut, then twist the extra bare wire around the ground screw and tighten it. If your inspector's a stickler, he may insist that you use "greenies," special wire nuts with an "out" hole on top. You slip the greenie over the new length of bare wire, twist it onto the splice so that the bare wire comes through the top of the wire nut, and *then* connect the bare wire to the ground screw.

Tails

When you're running wire from box to box, leave good long tails dangling from the boxes—at least 18", maybe even more. Later, when you're making the connections, you'll appreciate the latitude. And when you do make the connections, strip the wire so that the outer sheathing enters the box intact. It's only inside the box that you can have unsheathed wire. A good hard look at some properly wired boxes is worth all the prose in the world.

Junction Box

I recommend installing an interior junction box to contain the junction between the feeder cable from the house and the circuits of the outbuilding. Use a big square one to satisfy code and give yourself room for all the wire nuts. When you're finishing up the inside of the building, you can get a plain cover plate (with no holes for outlets or switch levers) to cover the box.

Feeder Cable

If the outbuilding is more than a stone's toss (a toss is shorter than a throw) from the house, 12-gauge UF cable won't suffice as feeder cable. It will probably be safe enough (assuming you put the whole outbuilding on a 20-amp breaker), but there'll be too much voltage loss along the cable. Consult this table as a guide to choosing the gauge of your feeder cable:

Distance between house and outbuilding	*Gauge of 20A feeder cable*
Up to 20 ft.	#12
Up to 40 ft.	#10
Up to 75 ft.	#8
Up to 125 ft.	#6

The table is conservative in that it assumes there'll be additional wire runs inside the house and outbuilding; you can't realistically expect to run wire as the crow flies, i.e., from nearest house corner to nearest outbuilding corner. You'll find, however, that heavy-gauge UF cable starts getting expensive out of proportion to its size. So you might want to check with your inspector to see if it's all right to run two or even three lengths of #12 cable in parallel. (Since it's probably the most common gauge of house wire, #12 is manufactured in massive amounts, which tends to keep the price low.) To run feeder wires in parallel, you connect the "hot" terminal in the house to the "hot" terminal in the outbuilding with the black wire in one length of cable, then connect the very same terminals all over again with the black wire in a second length of cable and perhaps still again with a third length. This distributes the current-carrying load among the parallel wires, which in turn reduces voltage loss along the wire. You may also want to consult with your inspector if your outbuilding is *really* far from your house or if it's on a breaker rated at more than 20 amps.

You can run feeder cable from your house to the outbuilding either overhead or underground. If you prefer to run it overhead (perhaps using trees to hold it up), consider the following points from the national electrical code, all of which apply to overhead feeder lines carrying *less than 600 volts*:

1. For overhead spans of up to 50', your feeder cable has to be #10 or thicker (i.e., #8, #6, etc.).
2. For any span exceeding 40', you have to string a "messenger" wire along with your feeder cable. Messenger wires don't carry current; their only function is to help hold up the feeder cable. Since they complicate the installation, I suggest that you keep all your overhead spans under 40'.
3. Where your cable passes over a roof, it has to clear it by at least 3'.
4. Otherwise, the cable has to be at least 10' above grade (or any porches, decks, or platforms) along the whole run. If it runs over a driveway, it has to clear the driveway by 12'.
5. The feeder cable has to leave your house *and* enter the outbuilding at a point at least 10' above ground. You also have to use "approved fittings" at the exit and entry points. Approved fittings are comparatively cheap and readily available at electrical supply houses. Tell the counter person what they're for, and (in most cases) you'll be given what you need. And while you're getting approved fittings, get some to attach the cable to the trees.

If you run your feeder cable underground, don't use anything smaller (i.e., with a bigger number) than #12. Beyond that, consult the table. National code requires that underground cable be buried at least 2' below grade if it's not encased in rigid metal conduit, which you certainly won't be in the mood to do. (It can be as shallow as 6" if it's in conduit.) But the code makes an exception of underground lines on a breaker of 30 amps or less. As long as you stay below 30A you can bury the cable to a depth of only 12", which is fairly easy to do without a power trencher. Still, think about the possibility of someone's using a rototiller or similar device near your cable. If it's not buried deeply enough, the results could range anywhere from interesting to catastrophic.

Assuming you won't be using conduit for the whole length of your feeder cable, code nonetheless requires that your cable make the vertical trip from underground up to your outbuilding (also down from your house if it exits above grade) inside a length of rigid metal conduit. However, now we're talking only about a couple of feet of conduit, so you may as well comply. You can have the conduit enter the outbuilding from underneath through an entry hole drilled down through the sole plate of the wall in which you put your main junction box. And after you install the feeder cable, seal around the hole with caulking compound. Also make sure to clamp the conduit to the junction box. The proper fasteners are commonly available at hardware stores and electrical supply houses.

Stringing Cable

When you're stringing cable inside the outbuilding, don't pull it extra tight between the holes you've drilled in the studs. If you plan to insulate, you'll want room to get the insulation between the wire and the siding.

Most inspectors like you to be generous with cable staples. ("Cable Staples" is what they're called on the box; the box also tells you what size cable they're for.) Where a length of cable runs horizontally from stud hole to stud hole, you don't have to staple it. But where it runs vertically along a stud, hammer in a staple every 18" or so. And staple it near where it enters every box. Since plastic boxes don't have cable clamps, you want something to hold the cable in place other than the connection itself.

Outdoor Switches and Receptacles and Ground-Fault Interrupters

Make sure to use only weathertight outdoor boxes for all exterior outlets, lights, and switches. Most good hardware stores have a display of outdoor electrical hardware. As long as the cable to the boxes goes directly through a siding hole and into the box (without an outdoor run) it can be ordinary indoor NM grade. And, for safety's sake (also to satisfy national code), have one of your outdoor receptacles be the kind with a ground-fault interrupter (commonly called a GFI). You install them just like ordinary outlet fixtures, but they give additional protection against accidental shock if you happen to touch anything "hot" while standing on wet ground; they're expensive but worth it. Like ordinary circuit breakers, GFIs trip at a given number of amps, so make sure they're rated for a slightly smaller current load than the breaker for the outbuilding itself. You want the GFI to trip before the fuse or circuit breaker does. If you're putting the whole outbuilding on a 20A breaker, get a 15A GFI. And once you've got a GFI-equipped outlet in place, it can protect additional receptacles too. Any outlet boxes wired down line from the GFI will trip it in the event of ground fault.

Circuit Breaker

Unless you're sure you know what you're doing, installing a new breaker in your house's breaker panel (or a new fuse in its fuse panel) is one job for which you'd best get help, either professional or expert-amateur or, at the very least, a book that's clear, reliable, and specifically applicable to the kind of panel in your house. Fuse and breaker panels come in many types, sizes, and incompatible brands. Generally one manufacturer's breaker won't snap into another manufacturer's breaker panel. And, if you do decide to hook up your feeder cable by yourself, by all means leave the job for last. That way you can fool around with the outbuilding wiring to your heart's content without danger to yourself or anyone else. Also, while you're working with the panel, make absolutely sure the main house fuse or breaker is OFF!

Ground Rod

Even if your house already has a working ground rod, put another one near the outbuilding and run some heavy copper wire from it to one of your ground screws. All the ground terminals in the outbuilding will be connected together, so any ground screw is as good as the next. Then, if something goes electrically afoul, the rampaging electrons won't have to go all the way back to your house in their frenzied quest for ground. You can buy a copper ground rod made just for the purpose, or you can sink a piece of metal pipe a few feet into the ground. Either way, make sure the connection to the ground rod is a good one. Metal pipe often has a thin skin of insulating corrosion on it, so sand it bright before connecting any wires and consider using solder. And as an added precaution before powering up the outbuilding, you can verify the accuracy of your ground wiring with a continuity tester. Clip one probe of the tester to your ground rod and touch the other probe to the ground terminals in all your outlets and fixtures. If you have perfect continuity in each case, you're in good shape. If you don't, fix the wiring so you do.

Phone and Television Cable Hookup

Telephone and TV-cable hardware is marketed in so many shapes and sizes that it would take a book-length guide to cover all the different kinds of jacks, plugs, and cables you can buy. My advice is to browse the displays at a good electronics store and consult with a knowledgeable salesman. Telephone cable virtually always comes with an even number of conductors in it. Each pair of conductors is called—not surprisingly—a "pair." The most common kind of cable—the kind sold for most do-it-yourself applications—is the two-pair type, which can accommodate two separate lines. But if you can find cable with still more pairs in it, and it isn't too much more expensive than ordinary two-pair cable, consider buying it. An extra pair may come in handy for an intercom or some signaling device. Like power cable, telephone and TV cable experience signal loss, so buy it in a gauge that's appropriate to the length of your run. Also like power cable, telephone and TV cable come with different types of plastic sheathing. If you plan to bury yours, make sure the sheathing is rated accordingly. For further hookup details, refer to the instructions on the back of the plastic-to-cardboard sandwiches that telephone and TV hardware is packaged in. Hooking up telephone and TV cable jacks is far less critical than house wiring. If you make a mistake you may get bad phone connections or unacceptable TV reception, but the voltages are too low to present any danger to life or limb.

Obviously the collection of points we've touched on constitutes more of a haphazard refresher course than a primer. Within limitations of space and patience, things couldn't really be otherwise. But if you're sure you're up to it, you can now go ahead and design your circuit, buy the materials, string the wire, wire the fixtures, and screw them in. And wherever you aren't clear about what to do or you sense there's any threat to your safety, get help. Avoid having it be said of you that you fell victim to unreasonable extremes of self-reliance.

13

Insulation and Vapor Barrier

It's at about this distance from the finish line, as the law of diminishingly visible returns sets in, that most professional builders start craning dewy-eyed toward their next project. Until now the fruits of your labors have been right up front for everyone to see, and for the most part the fruits in question have been watermelons. Now, disconcertingly, they start becoming raisins. You gaze at the outbuilding against the backdrop of an iridescent sunset, and you know the picture won't look any different if the building's insulated or not, trimmed or not, floored or not, heated or not. With the lion's share of the work done, the rest begins to seem like so much lily-gilding, even like—you hesitate before the phrase—home improvement.

Yet nowhere more than now do you need to call upon the stuff you're really made of. Almost anyone can fall prey to an attack of Emersonian machismo and enlist among the lonely ranks of those who'd dare put up a whole new structure. It's only a courageous few who have the deeper-running pluck to join the tract-house multitudes who hoard their days off to tackle do-it-yourself projects.

And installing insulation is the do-it-yourself project *par excellence*. The ins and outs of the material are so transparent, the techniques it calls for so accessible, that insulating a building is almost like striking a blow for democracy itself. When you go down to the local home center to take advantage of a sale on fiberglass batts, you thumb your nose at just the kinds of social stratification that the American and French revolutions sought to eradicate. Where (technology permitting) some effete aristocrat, shivering in his chateau, might have passed down orders for some luckless peasant to cart over a load of insulation and shove it into the palatial crannies only to itch like blazes for a week thereafter, contemporary democracy affords you the hard-won privilege of itching in your very own behalf.

Thomas Jefferson would be delighted. Himself an unabashed home-improvement enthusiast (and an abashed aristocrat to boot), he urged eternal vigilance against the reemergence of Old World aristocratic ideals. He knew how appealing those ideals could be, even if he underestimated their tenacity. Like certain viruses, the aristocratic ideal has been able to survive through sheer mutation. Just when one strain appears to be vaccinated out of existence, another, more resistant one pops up to take its place. Our modern strain, democratized but no less virulent, finds its vulnerable population in the burgeoning white-collar sector. There may not be any great proliferation of drafty old chateaux, but there are plenty of parvenu aristocrats—nowadays they like to be called "busy executives"—who will get on a tear about shaving fuel consumption and make phone calls to insulation contractors. The contractor will send over a laborer—maybe not a luckless peasant, but close enough—who puts on goggles and a mask, unrolls the wads of mineral fiber he's brought along with him, cuts them into strips, stuffs them wherever they're supposed to go, and, because he can take a shower and leave his work clothes in a washing machine, itches only for a few hours. Meanwhile the executive, who prefers to limit his direct participation in matters of home improvement to making key décor decisions, has kept his hands completely clean. He's spoken on the phone; presently he'll sign a check. He hasn't put his body on the line because he's saving it for racquetball and golf, activities not tainted by any obvious usefulness.

But you, dear reader, are presumably cut from different cloth—not the tightly woven sharkskin of the boardroom but a simpler, sturdier broadcloth reminiscent of colonial America. You're a doer, not just an intermediary. When basic needs arise, you like to see to them yourself. Not for you the uneasy, even twitchy arrogance of all those parvenu aristocrats. (Why shouldn't they be twitchy? Unlike their Old World counterparts, they lack the old support systems. If the shivering marquis's chateau was still drafty, he could have his insulator flogged. If the executive's fuel bills don't come down, about the best he can do is ventilate impotently to some consumer protection agency or start a suit.) You've read in one or another magazine that privately, in the sanctuary of their analysts' offices, many of today's self-styled aristocrats complain of feeling somehow like frauds, and, now that you're nearly done with the outbuilding, you aren't so surprised. It's probably the going price for placing greater value on nebulous credentials than tangible accomplishments.

No such fraudulence for us, however. If contemporary man has surrendered some vital portion of himself in trying to adapt to modern life, nothing says it can't be gotten back. And if doing things yourself feels like a step in that direction, perhaps the tract-house multitudes are on to something after all.

So let's proceed with your redemption by insulating the roof. Your rafters are all 2x6s, and all the roof bays are more or less $14\frac{1}{2}''$ wide. The indicated type of insulation for this purpose is R-19 fiberglass. (R value is a measure of insulating power. The greater the R number, the more effective the insulation.) It comes in tightly bundled $14\frac{1}{2}''$ rolls, and when you open the bundles it expands to just the right thickness to fill the $5\frac{1}{2}''$ space that 2x6s give you. All you need to do is cut it into strips of proper length, and virtually any sharp knife will serve the purpose. Calculate the interior square footage of your roof (I say "interior" because the inside area of the roof—actually the ceiling—is somewhat smaller than its outside area, and you don't want to cram insulation into the eaves lest you plug up the soffit vents), then buy that much R-19. Bundles of insulation are plainly marked with their square footage.

If your dealer stocks many different types of insulation, you'll notice that it comes in precut batts as well as continuous rolls. However, it's unlikely that any of the standard batt lengths will conform to the interior length of your rafters, so buy continuous rolls. You'll also notice that insulation can be had unfaced or faced with foil, kraft paper, asphalt felt, clear plastic, and even combinations of these facings. Unfaced insulation is intended for friction-fitting. You stuff it between a pair of studs or joists, and the friction between the framing members and the fiberglass holds the insulation in place. Normally that's fine for walls, but for ceilings it's a little chancy. A sudden gravity surge could have you spitting fiberglass for days. I'd therefore recommend that the R-19 you get be faced with *something*. (I also recommend that you wear gloves, a fiber mask—if you don't feel like spending fifty cents at the hardware store, at least tie a handkerchief over your nose and mouth—and plastic goggles; fiberglass particles aren't toxic, but they're an irritant to eyes and mucous membranes.) Any facing, regardless of what it's made of, has a running fold-over tab on either side that will let you staple the insulation to the rafters.

Apart from getting purchase for your staples, though, the question of which facing to choose remains a muddy one. The oil shortage of the 1970s gave sudden currency to the subject of energy conservation, but, as technologies go, it's always been a little on the soft side. Foil-faced insulation is supposed to reflect radiant heat back into an enclosure; it probably does, but many experts say it's not enough to make much difference. Paper-faced insulation is more modest in its claims: it gives you something to put staples through and keeps the fiberglass from showering dust.

Vapor Barrier

Insulation with plastic or asphalt-felt facing purports to give you a vapor barrier, but unless it's stapled with exquisite care, it's doubtful how effective such a barrier can really be. The thing is, when you heat a room on a cold day, the inside of the insulation becomes considerably warmer than the outside. As normally moist room-temperature air meanders outward through the insulation, the water vapor it contains can begin condensing on the colder insulation fibers and even onto the studs and siding. If condensation accumulates, it can promote rotting within the walls. The purpose of a vapor barrier is therefore to keep interior air from sifting into the walls in the first place. Now, although metal foil and sheet plastic are functionally impermeable to vapor, there remains the matter of the seams between the strips of insulation—one for every bay. It's true that you could find some way to seal them, but that's likely to be much more trouble than it's worth. Alternatively, you can let the insulation serve only as insulation and install a separate vapor barrier, just as you did when you built the deck. Any plastic that's transparent and at least 4 mils thick will do the job.

While we're on the subject, you may be wondering why the interior walls themselves aren't vapor barrier enough. Well, in some cases they are. Plywood panel walls tend to leak vapor through the seams between the sheets, but Sheetrock walls that are decently spackled, primed, and painted (either two coats of primer and one of paint or one of primer and two of paint) are reasonably vapor-tight. You can even get primer that's specifically rated as a vapor barrier, in which case a single prime coat may be enough. Ask about it at the paint store and then just follow the directions on the can.

Ultimately the entire question of what to do about vapor barrier is another judgment call that largely boils down to how damp you think the inside of the outbuilding is likely to get on cold days. If you plan to use the building summers only, vapor barrier needn't concern you. In winter, with the unheated interior at pretty much the same temperature as the outdoors, condensation simply won't occur. Nor will it be much of a problem if you keep the building plumbing-free. (And most townships look fiercely askance at plumbing in an outbuilding; they're afraid you'll make it a hotel.) Occasional evaporation from vagrant coffee cups and the surface of your skin is hardly likely to rot out the framing during your lifetime. Yes, the illustration shows a separate plastic vapor barrier over all the insulation, but that's because the outbuilding we've been following in pictures was built (in part) to be a source of illustrations.

In any event, now that you've got your R-19 tucked between the rafters, you can fit some R-11 between the studs. The expanded thickness of R-11 fiberglass matches 2x4 stud walls in

Vapor barrier over insulation

the same way that R-19 matches 2x6s, and, like R-19, it comes in $14\frac{1}{2}''$ rolls or batts, each with a choice of facings. My own preference is for unfaced $92\frac{5}{8}''$ batts. If you've been using precut studs, one batt will fit nicely into every full-sized bay without any cutting or stapling. You'll probably be shocked at how few full-sized bays you actually have, but by using batts you'll still keep cutting to a minimum. And if you do cut strips from faced rolls, you can probably skip stapling them. Stuffed between wall studs, even faced insulation usually gives a satisfactory friction fit.

For good measure, pull off some loose wads of fiberglass and stick it into the crevices between the door and window casings and the rough openings that surround them. Don't cram it in too tightly, though. Fiberglass insulation works best when it's fully expanded. The more you compress it, the less efficient it becomes.

Now, if you've decided in favor of a separate plastic vapor barrier, go ahead and tack it on. And since you're going to the trouble, try not to puncture it with more staples than you need. Once the interior walls are up they'll hold the plastic nicely in place.

And there you have it. With all that insulation in the walls, there'll be days when you'll almost be able to heat the building with a candle. On long winter nights, with the windows shut (except for one that's open just a crack) and no one around to tend the wood stove, it will retain a fair amount of heat until you arrive to stoke the next day's fire. And with a careful vapor barrier in place, the building may become so airtight that when three or four people inhale at the same time you'll run the risk of sucking birds in through an open window.

14

The Interior Walls

PANELING OR SHEETROCK?

If you'll be installing wallboard in the outbuilding, you've got an interesting and difficult choice to make. Which do you go with, paneling or Sheetrock?

All right, maybe not so difficult or interesting a choice if you already love one and hate the other. If it's as clear as that, install the one you love and shun the one you hate. But if it isn't all that clear, you might want to pause for a moment to contemplate the relative merits of each.

To start with, Sheetrock, after it's done and painted, is the more expensive of the two, but not so much so as to tilt the decision very sharply toward paneling.

Paneling may also be the more congenial to your own personal notion of what an outbuilding ought to be. It evokes some of the atmosphere of rec rooms and finished basements, but it's perfectly conceivable that you're building an outbuilding because you want an above-ground finished basement. Sheetrock makes a room feel like a real room, which is fine if that's what you want and maybe a touch too formal if it's not.

Both paneling and Sheetrock come in 4x8 sheets. Paneling is physically light; Sheetrock is heavy. Paneling is easy to handle; Sheetrock isn't. Sheets of Sheetrock can break of their own weight if you don't support them carefully while you're carrying them. If you drop one on its corner, you can expect to smash the corner.

Paneling doesn't have to be taped and spackled after it's nailed up. Sheetrock does, and taping and spackling require patience, concentration, and a good amount of time. Then again, paneled walls nearly always have to be trimmed with molding where they form a corner with another wall or with a ceiling; Sheetrock walls don't. But mitering and nailing molding take only a fraction of the effort that it takes to tape and spackle Sheetrock.

If you find a kind of paneling whose surface you're happy with (it comes in hundreds of colors and textures), you don't have to paint it; new Sheetrock needs at least two coats, one of primer and a final coat on top.

Paneling is easy to nail to ceilings; Sheetrock isn't. But paneling has to be cut with a saw, while all you do with Sheetrock is score it with a Sheetrock knife and then just break it off.

Paneling is thin and feels it. When you tap it, it drums. Sheetrock has a more substantial feel. And paneling doesn't give you very airtight seams between adjacent sheets; Sheetrock does. When Sheetrock is decently taped, spackled, primed, and painted, it looks and feels a lot like painted plaster. Paneling looks and feels like—well, like paneling.

Paneling: Easy

On balance, paneled walls are vastly easier to do. To take you step by step through each and every detail of their installation would be to belabor the obvious and then some. You put a plywood blade in your circular saw, you figure out how many 4x8 sheets you'll need to cover your walls and ceiling, you buy the paneling, you get a box of paneling nails (which even come in a variety of colors to match the paneling itself), and you go to work. Even more than fiberglass insulation, paneling has the power to send inveterate do-it-yourselfers into their own personal version of pig heaven. Installing it barely qualifies as labor; it's more like occupational therapy on a grander scale than lanyard weaving.

Moreover, you can transfer most of what you learned when

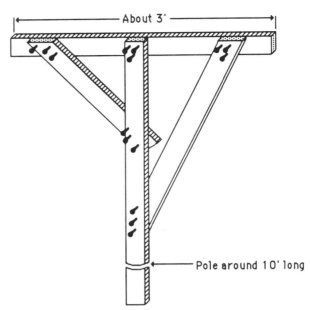

A crude but serviceable wood panel and Sheetrock holder-upper

you installed T-111 on the outside of the building. Without bothering to worry too much about the neatness of the cuts along the corners (they'll be concealed by corner molding) or the cuts around the window casings (they'll be covered by window trim) or any gaps between the bottoms of the sheets and the floor (they'll be covered by the baseboard), you measure it, saw it, and then just nail it up. The only cuts that require close attention are the places in the ceiling where the tie beams come through and the openings in the walls for electrical boxes, and in the latter case you can hide your mistakes by buying extra-large wall plates for the switches and receptacles. You'll probably have to make a T to support the ends of the ceiling panels when you nail them, just as you will if you're using Sheetrock, but apart from that the whole thing's as easy as proverbial pie. Once the paneling's nailed up, you go to the lumberyard and pick whatever corner molding pleases you, you stain or paint it if you're so inclined, you nail it into all the corners with finishing nails, and you're pretty much done. About the only piece of molding that might present any difficulty is the broad strip that trims the ceiling peak just under the roof ridge, but we'll talk more about that in the next chapter.

After you're done, you can assess the look of it, and if you don't like the color or the texture that it came in you can paint right over it. White paint has a way of obliterating all manner of annoyances. In the last analysis, the only real problem that paneling is likely to present you with is that, painted or not, it stubbornly insists on looking and feeling like paneling. But if you can't see why that should strike you as a problem, drop the whole issue and go on to the next chapter.

Sheetrock: My Choice

I happen to prefer Sheetrock. I was willing to take the extra trouble to install it, and I'm prepared to admit that my preference may be totally irrational. In fact, halfway through the project I began to regret the decision. I was ready to stop building the outbuilding and start using it, and if I'd gone with paneling I'd have been long done. Yet in the end I was glad I went with Sheetrock. This book got written in the outbuilding, and when I'm writing I like to be in a room that feels like a room, not the inside of a gigantic packing crate. Sheetrock gave me the illusion of civilization I needed.

In order to penetrate the Sheetrock barrier, however, you'll have to harness a few new skills, most particularly the taping, spackling, and polishing of raw Sheetrock. Notice that I've again been careful to say "harness" and not "master." Unless you're a seasoned spackler, you'll be practicing as you go along, and it's inevitable that the outbuilding will wind up bearing some of the scars of your experiments. So if you believe that finishing an outbuilding demands every bit as much perfection as finishing, say, a formal living room, now's the time to change your mind. Many people are convinced that crafts like spackling (I'm talking about spackling whole rooms, not an occasional crack) are best left to experienced veterans, but I suggest you try being more optimistic—justifiably or not—about what you're capable of doing. Once you're finished you may come to realize that your optimism was totally unfounded, but half the fun of projects like these is their ragged edge of indeterminacy.

Actually, the notion that certain skills should be performed exclusively by certified craftsmen can undermine the novice builder's confidence at just about any stage of construction, and for that reason alone it deserves a closer look. I can't say what its earliest origins are, but it was widely propagated by the medieval craft guilds as a means of self-protection. You went through a formal apprenticeship, you gained admission to a guild, and only then were you qualified to practice your craft. The necessary skills were simply too numerous and complex for most ordinary folk. Once you'd mastered them you kept them to yourself, even to the point of shrouding them in mystery. So long as the mysteries were accessible only to you and your fellow guild members, business was apt to remain good, and in the event you needed it, you even had some handy leverage over the odd aristocrat. The more powerful the sense of mystery, the more indispensable you'd be.

Eventually it was that very institutionalization of mystery that helped give rise to all manner of occult fraternal societies. During the Middle Ages the Freemasons came into being as a select group of stoneworkers, and what began as a collection of closely guarded secrets of the mason's craft—recipes for mortar and the like—evolved over time into the whole Masonic alchemy of symbols and ciphers that would mark the entrance to a much larger body of arcane "wisdom." Back then the mysteries inside your walls and the mysteries at the heart of man's existence weren't seen as such very different things.

But then the spirit of the Age of Reason took hold, and with it came a new impulse toward demystification. No one better typifies the transition than Benjamin Franklin. Known to countless magazine readers as a spokesman for the ancient mysteries of the Rosicrucians (the Masonic-style fraternity of which he was an active member), he was a skilled craftsman and businessman—a printer by trade—who was unafraid to venture into just about any area that caught his interest. Unlike Thomas Jefferson, whose origins among the Virginia gentry gave him tacit permission to dabble, Franklin's working-class beginnings might normally have been expected to keep his activities inside a fairly narrow set of constraints. But the times were bubbling with possibility, and he went on to become an amateur physicist, writer, experimenter, inventor, politician, revolutionist, diplomat, and rake who never felt much awe for the invisible boundaries that surround most new endeavors.

In a slightly less momentous way, it's the mission of this book to encourage a little of the same lack of awe for the invisible boundaries that surround some of the building crafts. In most cases, the only things you really need to vault those boundaries are a sense of competence, a little ingenuity, and a measure of venturesomeness. You proceed as though it never occurred to you that spackling can't be done by anyone but spacklers, and you simply go ahead and spackle. Then, while you're spackling, you needn't be encumbered by any ambition to become a pro-quality spackler because it isn't the credential you want, it's a set of decent-looking walls. And, that being so, let's get on with them.

SHEETROCK

Standard Sheetrock is manufactured in thicknesses of $\frac{3}{8}''$, $\frac{1}{2}''$, and $\frac{5}{8}''$, and the difference in the way each thickness feels is greater than the mere $\frac{1}{8}''$ gradations would suggest. The $\frac{1}{2}''$ is by far the commonest. Builders use $\frac{5}{8}''$ when they want a "luxury" feel and $\frac{3}{8}''$ when they want to save time and money. I suggest you save time and money and use $\frac{3}{8}''$. You're dealing exclusively with outside walls (i.e., walls that separate the room from the outdoors as opposed to walls that partition one room from another), so you're not concerned about between-room noise. And if you're doing Sheetrock walls you've probably already put in insulation, which will muffle the little bit of extra drumming that you get from $\frac{3}{8}''$ as compared to $\frac{1}{2}''$.

Installing Sheetrock

Use Sheetrock nails about $1\frac{1}{4}''$ or so long (and do use Sheetrock nails; common nails won't work anywhere nearly as well). Space the nails about 1' apart along the studs and rafters. When you hammer the nails, try sinking the nailheads a fraction of an inch below the surface of the Sheetrock without tearing the paper coating. The shallow well that your hammerhead will make in the Sheetrock around each of the nailheads is called a "dimple," and later, when you fill the dimple with Spackle you'll conceal the nailhead.

You can start either with the ceiling or the walls, but whichever one you choose, finish that part before going on to the other, so you have a uniform joint between your sloped ceiling and the walls. Although it's common practice to nail up Sheetrock with the 8' edges horizontal, I'd suggest you adopt the *un*common practice of keeping the 8' edges vertical. Where very large unfenestrated walls are being Sheetrocked, common practice gives you a big, wall-length horizontal seam that's at a comfortable 4' height for taping and spackling. But in the outbuilding that isn't an especially important consideration, and putting the sheets up vertically will probably save you a good amount of cutting. When you do put them up, make sure to tear away all the paper strips around the edges (the ones that hold the new sheets together in pairs). And if the ceiling's already "rocked," butt the top edges of the sheets tightly against the bottom edges of the ceiling sheets so that you won't have to fill any yawning gaps with wallboard compound. (I'm using the brand name "Spackle" and the term "wallboard compound" interchangeably, just as they're used in real life.) If you're working alone, keep assorted wood scraps handy so you can wedge them under the bottoms of the sheets until you nail them to the studs. Nice clean butt joints along the tops of the sheets may leave you with a few inches between their bottom edges and the subfloor, but that's one of the reasons that God created baseboards.

As I said before, you cut Sheetrock by scoring the "good" side with a Sheetrock knife (cheap to buy at hardware stores), then snapping it where it's been scored, and finally cutting through the paper on the other side with the same knife. If you can arrange to have a Sheetrock square around, you'll find it remarkably handy for drawing lines and also for guiding the knife blade when you make the cuts. A Sheetrock square is essentially a 4" T-square, and it supplies you with lots of straight lines and right angles when you need them most.

As with paneling, the border cuts around windows and doors

"Rocking"

needn't be perfect, since you'll be nailing some kind of wood trim over the junctions. Where you want to be careful, though, is with the cutouts for the electrical boxes and with any cuts that butt against more Sheetrock. Try to arrange things so that all the interior joints in any wall or ceiling surface are formed by factory edges. Leave *your* cuts for the perimeter of the wall, where they'll be smoothed over by all the wallboard compound that you'll trowel into corner joints. Where one factory edge

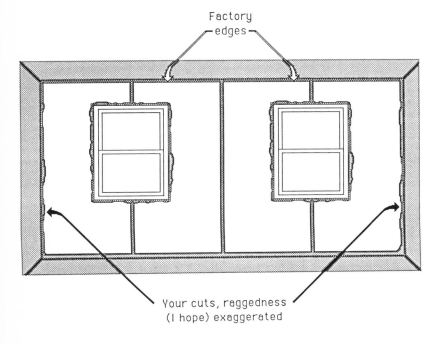

butts another, you'll be able to take advantage of the indentations along the edges of the Sheetrock to ensure that every wall and ceiling surface is a single, flawless, uniform expanse.

Two more points we should touch on before you start "rocking" are nailing Sheetrock to overhead surfaces and making patches. On the matter of overhead Sheetrocking, I strongly suggest that you (1) use a homemade holder-upper like the one pictured earlier in the chapter, and (2) get help. Yes, you can always make two or three such T-braces and maybe succeed in rocking the ceiling by yourself, but it's far easier to have one person support a sheet while the other person squares it up and nails it. As for patches, arrange to have as much nail bearing as possible around the perimeter of each of the holes you're cutting a patch for, even if it means you have to carve up some of the surrounding Sheetrock. Of course, when you tape and spackle an inserted patch, you seldom have the luxury of factory indentations to embed the tape and wallboard compound in, so you'll have to do a little extra sculpting with your trowel. To keep the sculpting to a minimum, though, cut your patches carefully. Wallboard compound is made to fill narrow seams, not navigable canals.

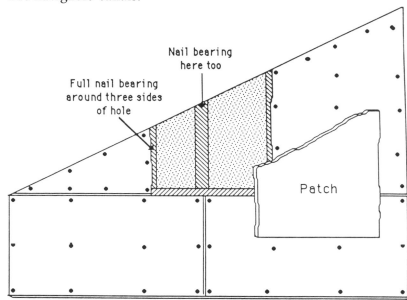

Sheetrock patching

Taping and Spackling

For taping and spackling, you'll need a roll of Sheetrock tape, a can of Spackle or similar wallboard compound, and three tools: a putty knife with a 3″ or 4″ blade (that's 3″ or 4″ *across*), a big spackling trowel about 1′ across, and a corner trowel. All Sheetrock tape used to be made of paper, but now there's a fiberglass version that's ever so slightly easier to work with. Both kinds are cheap, and the differences are matters of nuance.

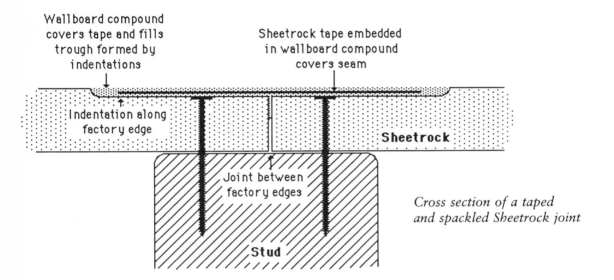

Cross section of a taped and spackled Sheetrock joint

Wallboard compound—Spackle—can be purchased as a powder that you mix with water or as a premixed paste. The premixed kind isn't that much more expensive, and I strongly recommend it. The tools are pretty rudimentary, they're stocked by most hardware stores, and they're available in several grades of quality (and price). Since you'll probably be doing this just once, there's little point in buying high-end tools that will lie around corroding in the coming years. Expensive trowels are for hard-core spacklers, and I'm guessing that your overarching aspirations lie elsewhere. On the other hand, really cheap trowels don't do a very good job, so it's mid-range tools you'll probably want to buy.

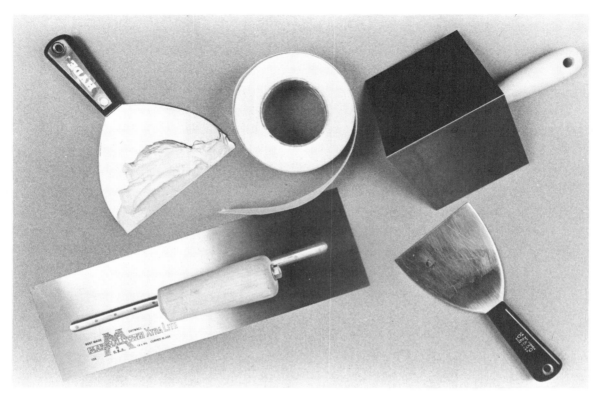

Spackling tools

To gain a feel for the process, start by picking a good long seam between two factory edges. Capture a blob of Spackle on the big trowel (the straight one) and dab it into the shallow trough formed by the depressions along the edges of the sheets. Keeping your trowel strokes as long and fluid as you can, fill the trough with a continuous ribbon of Spackle. Don't worry too much about smoothness yet. The Spackle will remain workable for a while, and you can smooth your wall off later. Now tear off a length of Sheetrock tape and press it gently onto the wet Spackle. Try to keep the tape stretched taut, and center it over the seam as best you can. The Spackle will hold it temporarily in place.

Then, replenishing the Spackle on your trowel as you use it up, run the trowel edge smoothly along the tape. Press down firmly as you move the trowel so that (1) the tape gets embedded in the Spackle beneath it, (2) a thin layer of Spackle gets deposited on *top* of the tape, (3) the whole mess gets pushed into the trough along the seam, and (4) the stuff in the trough is level with the surface of the Sheetrock. It's true you're being asked to satisfy four separate conditions with each trowel stroke, but you shouldn't find it all that hard to do. And if you notice any imperfections, trowel them away. You can make as many passes over the seam as you like.

Incidentally, you don't have to cover the tape so perfectly with Spackle as to render it completely invisible. Spackle becomes a lot more opaque when it dries, and farther down the road, a coat of primer followed by one of paint will conceal a multitude of sins. Besides, you'll get another chance to pretty up the seam after it's dry. Raw walls are normally given two coats of Spackle—a base coat and a polish coat—since dry Spackle reveals its flaws more readily than wet.

Once you've finished the seam, go on to another easy one and then another. Finish all the inside seams before getting to the corners. That way you'll have a chance to become familiar with the tools and materials, and by the time you have to use the corner trowel you'll feel more comfortable with what you're about to do.

Eventually you'll run out of factory-edge seams and be forced to deal with cruder ones. There's every likelihood that some of the seams in the triangles that top off both your short walls won't have factory depressions along their lengths, but by now you shouldn't have much trouble handling them. You prime the seams with Spackle, always making sure that there's enough of it to glue down the tape. (Regardless of the kind of seam you're doing, the key to taping Sheetrock is laying down enough Spackle to glue every last square centimeter of tape to the Sheetrock. A poor job of taping can masquerade as a good one for a while, but three years later, when it starts to peel, you'll wish you'd been more careful.) After the seam is primed with Spackle, you stick on the tape and trowel another coat of Spackle over it. Without those troughs to work with, you have to attend more closely to feathering the edges of the Spackle. (In case you don't know, "feathering" is a term from masonry that refers to the gradual tapering of the thickness of a patch so that it merges smoothly into the surface of the material being patched.) But a spackling trowel is a dandy feathering device, Spackle readily succumbs to feathering, and you can always touch up the seam when you do the polish coat.

After you've finished all the flat seams (i.e., the non-corners), spackle the remaining nail dimples with the putty knife. Just put a little Spackle near the edge of the knife, dab some into each dimple, then smooth it off with one fluid stroke. You'll soon find that it takes even less time to spackle Sheetrock nails than it does to hammer them in.

Now try a corner formed by two walls. (Not being square, the corner formed by the top of the walls and the sloped ceiling is more difficult to do, so save it for the end. While you're still learning, it always makes sense to do the easy stuff first and then bring your new skills to the harder tasks ahead.) First dab a bunch of Spackle into the corner with the straight trowel, then use the corner trowel to smooth it along the seam. If there's any excess Spackle, leave it. Too much is better than too little, and you can always use the corner trowel to squeeze it out later. After you've laid down a base coat, tear off a length of Sheetrock tape, fold it in half lengthwise, and stick it gently along the corner so that half its width clings to one wall and half to the other. Then, taking care not to crimp the tape, press it smoothly into the Spackle with the corner trowel. You'll find that corners are more of a challenge than straight seams, but if you concentrate on what you're doing you'll get a feel for them in a surprisingly short time. Now, once more keeping your strokes long and fluid, use the corner trowel to (1) lay a thin top coat of Spackle over the tape, (2) squeeze excess Spackle from *under* the tape, (3) feather the edges of the Spackle, and (4) form a nice square corner in the wall. Yet again that's four things to be thinking about at the same time, but any number of people who'd never strike you as being able to hold *two* thoughts in their head at the same time have gone on to become master spacklers. Be thus encouraged.

When any seam is dry, it's ready for a polish coat, so you can be applying first coats to some seams at pretty much the same time that you're putting polish coats on others. Depending on the looks of a given seam, polishing may not even call for additional Spackle. A light sanding may be all you need. (Wear a dust mask when you sand Spackle; the vinyl particles are unfriendly to lungs.) You may even be able to get away with a few quick wipes with a damp sponge. Or you can sand and/or sponge and then respackle. It all depends on how careful you were the first time around. Most brands of wallboard compound advertise themselves as not shrinking when they dry, but that's seldom the case. The compound in your deeper nail dimples will probably have receded some, and you may want to touch them up. The same holds true for just about any deposits that have much depth to them. So inspect your work, patch what needs patching, polish what needs polishing, then look at your spackled walls and imagine them with a couple of coats of paint blending all the different shades of white and gray and with neat lengths of trim covering all the rough edges. You've already made an outbuilding. Now you're on your way toward making a *room*.

15

Completing the Room: Trim and Flooring

Trimming out a "raw" room—and we'll include the installation of flooring in the trimming process—calls for a somewhat different mind-set than the one you've probably been cultivating thus far. Falling, as it does, somewhere in the no-man's-land between construction carpentry and interior decoration, making trim asks that you exchange a measure of the swagger, bravura, and sheer muscle that you brought to the tasks covered in the preceding chapters for a greater degree of subtlety and even delicacy. You don't exactly have to be an artist when you trim a room, but you do have to be more than just an inspired laborer. Maybe what you need to be is an artisan. For the first time during construction of the outbuilding you find yourself paying close attention to the appearance of the wood—to grain and knots and cracks, to jagged sawblade tracks and peeling slivers. Most of what you've made so far has been covered up with something else. Trim, by contrast, is where the buck abruptly stops. If any one material is emblematic of the new mind-set I'm talking about, it's sandpaper. Until now you've kept a cozy distance from the stuff. While trimming out the room you inevitably consider using it.

To boot, if this is your debut as a trimmer you're about to do something mildly momentous. In the past, the look of every room in every house or apartment you've ever bought or rented has probably come to you as a given. The trim was already nailed on and most likely painted over. An alien sensibility had conceived the margins of your space—the borders of your windows, your floors, the edges of your walls. Now you're confronted with an empty canvas; you can have the room look the way *you* want it to look. If getting your own room when you were a kid was any sort of milestone, *making* your own room can be a bigger one.

A century ago, rooms were often trimmed in the style of baroque picture frames. Even if most large objects—houses, for example—were expensive, details were comparatively cheap to buy. Nowadays, large objects are still expensive, and details are altogether out of sight. So in modern rooms the bulk of the trim consists of the woodwork around the windows, probably because windows survive as something more than just undifferentiated fenestration. They still top the list of things to sit beside when you're seized with an unruly urge to daydream. Especially when you're deep in thought, a window puts the outside world at one remove. What it gives you is a picture of the world, not the world itself. The trim around the window frames the picture. Ignoring this, builders of new housing commonly seek to save money and labor by holding window trim to a bare minimum. They'll omit a lot of woodwork and try instead to bring as much Sheetrock as they can to the edges of the factory-built units. But I suggest you not breeze by the trimming of the windows as rapidly as that. In choosing the type and width of your window trim and the kind of wood you make it with, you'll have the chance to influence how the room will look for a long time to come. It's a chance that shouldn't be allowed to slip by.

Basic Trim Designs

The sketch on the next page shows the five component pieces that make up the basic interior trim of a window: the inside top casing, the two pieces of inside side casing, the stool, and the apron.

Although the three casing pieces and the apron are shown as being made from simple rectangular-cross-section stock, there's nothing to keep you from scouring the lumberyard for something fancier. Be aware, however, that clear (i.e., knot-free)

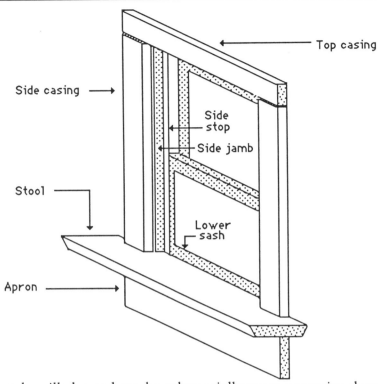

Window trim: the basic parts

ornately milled wood can be substantially more expensive than pieces of #2 (i.e., knotty) pine that you buy as 1x4s or 1x6s or rip down from larger boards yourself. I happen to like the look of plain #2 pine in a rustic outbuilding, especially if it's given a clear finish and otherwise left unpainted. The trim in the photographs at right was all home-ripped on a table saw from $\frac{3}{4}''$ pine shelving. You can even rip trim with an ordinary circular saw if you're very careful, or you can get one of those gadgets—they've gotten to be quite cheap—that lets you mount your circular saw so that it *becomes* a table saw. And if you have any facility with a router or you're willing to acquire it, after you rip down your trim you can use the router to custom-mill it to taste. You have, in other words, a multiplicity of options all along the spectrum between expensive-and-easy and cheap-but-labor-intensive. For the sake of simplicity we'll proceed on the assumption that you're using plain rectangular-cross-section stock. Should you prefer something more elaborate, you can easily enough extrapolate.

Window trim detail: top casing

Window trim detail: casing, stool, and apron

A fully trimmed window

The most reasonable way to begin is by making a few basic decisions. The first concerns paint. If you plan to paint the trim, trim out the room now and paint it afterward. The trimming process is bound to leave you with a residue of mental notes on all the minor errors you've committed, but you can console yourself with the knowledge that you'll soon be coming around again with crack filler, primer, and paint, and that most of the errors will disappear into the blend. If you plan *not* to paint the trim, first paint the room and *then* trim it. That way you won't have to mask the trim while you're painting, and the worst thing you'll have to worry about is soiling your new paint job, which can always be touched up with the flick of a brush anyway.

But there remains the matter of the floors and baseboards. Baseboards go in on top of the final layer of flooring, and this may be the only chance you'll ever get to paint the room without having to concern yourself with splattering the floors. So, whatever else you do, leave the floors and baseboards for last. It makes perfectly good sense to *cut* the baseboards while you're cutting the other trim, but hold back on nailing them down. With only beat-up subfloor underfoot, you can let the outbuilding be your shop while you're prettying it up, which means that you can drizzle paint and sawdust to your heart's content and not have to lay dropcloths wherever you're working. Later on, regardless of whether you've painted first and then trimmed or done it the other way around, you can floor over all the mess on your subfloor (a dramatic event when it actually takes place) and then nail down the baseboards and call it quits. If you've painted all the other trim, you simply paint the baseboards before nailing them, then touch up the hammer tracks with a small brush. It's the rare paint job that lets you slosh away so recklessly, so you might as well enjoy it while you can. (See the next chapter for some more remarks on paint.)

Still another basic decision to get out of the way now concerns the width of the inside side casings around the door and windows. (Doors are trimmed just like windows, except they obviously don't need stools and aprons.) For a feeling of stylistic unity, it's best to have all the side casings in a given room be of the same width, and a working minimum width can be obtained by going around with an ordinary ruler, examining the doors and windows, and finding the maximum distance between the inside edges of the door and window jambs and the edges of the Sheetrock or paneling that surrounds them. If the room is to be at all presentable, that maximum distance determines the minimum distance that the trim has to span in order to conceal the gaps around the jambs.

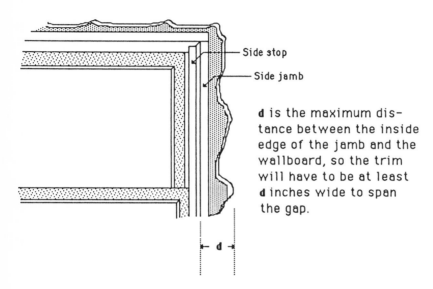

I don't think I have to point out that if **d** in the above sketch is much greater than the maximum side casing width that's acceptable to you, you'll do best to repair the wall and reduce **d**. There's no reason to have elephantine side casings on all the windows just because you've made one or two flagrant errors.

You'll find, however, that the top casings don't mind being a little wider than the side casings, and the aprons beneath the windows generally look best if they're *considerably* wider than the side casings. (Factory-made windows are usually constructed in a fashion that virtually demands that the aprons be fairly wide.) The best way to gain a sense of what you want is to look at a variety of windows in a variety of houses and then choose the style that suits you best.

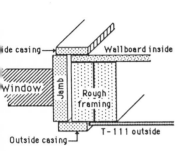

What you'd like to have is this. The inside surface of the jamb is flush with the wallboard, so you can nail the inside casing right across.

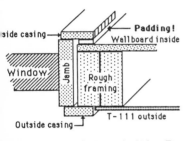

What you probably have is this. The inside surface of the jamb comes out past the wallboard, and you need to put a strip of padding under the inside casing to make everything come out flush.

Padding the trim

Padding the Trim

Another factor you'll have to account for is the very great likelihood that the edges of your jambs aren't flush with your walls, which means you'll have to do some improvisatory padding.

Window and door manufacturers can do no more than guess at how you'll make your walls, and most assume you'll be using $\frac{1}{2}''$ sheathing outside and $\frac{1}{2}''$ Sheetrock inside when they decide how wide they'll mill the jambs. That's *most* of them. A few of them make still other assumptions, and then there are the everyday irregularities that crop up in all building materials compounded by yet more irregularities in the way people *use* the materials. If you've put up $\frac{5}{8}''$ T-111 and $\frac{3}{8}''$ Sheetrock, you may find that you don't have to do any padding at all. If you've used wood paneling inside, it's almost certain that you'll have to pad your trim, and if you've chosen not to put up any kind of wallboard you won't be trimming your door or windows either, which means that you're excused from class for the rest of the day.

As for what to pad with, that depends on how much you have to pad. Don't expect the thickness of the padding to be consistent all around the room, though. One window may want $\frac{1}{2}''$ padding, while $\frac{3}{4}''$ may work best with another. The grim truth is that the majority of the padding will ask to come in oddball thicknesses in which no piece of wood has ever been milled. So if your standards call for perfect fit in every case, prepare either to revise your standards downward or shave an awful lot of wood.

Most of the trim padding for the outbuilding in the illustrations needed to be $\frac{3}{4}''$ (more or less) thick, so I simply ripped it down from $\frac{3}{4}''$ #2 pine shelving, which was the cheapest source of $\frac{3}{4}''$ padding I could think of. If you're not inclined to start ripping lumber, an assortment of preripped alternatives can be found at the lumberyard. One-by-two pine furring strips are $\frac{3}{4}''$ thick and $1\frac{1}{2}''$ wide, and they're also quite cheap. Furring strips are normally used as nailers for other boards, so they aren't apt to be gorgeous. You'll probably want to give a quick sanding to the edge that faces outward toward the room. For $\frac{1}{2}''$ padding there's always ordinary door stop (the stuff that's nailed around the insides of interior doorjambs to prevent doors from swinging past where they're supposed to swing). Most stop is smooth enough not to require further sanding, but it's more expensive than furring strip. Ultimately the best way to choose your padding is to ask the people at the lumberyard what they've got and what's cheapest. You're a respected figure there by now (another way of saying that you've run up a good-sized bill), so you can expect more cooperation every time you stop by.

Window Aprons

When you trim a window, the first thing you make is the apron. The apron should be wide enough to cover the hole between the top of the sill and the lower edge of the surrounding wallboard, and in most cases its length equals the distance between the outer edges of the inside side casings.

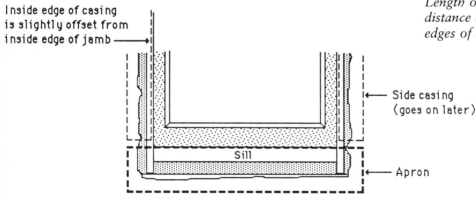

Length of apron equals distance between outside edges of side casings

Try to arrange that all the aprons in the room be of the same width (i.e., the same distance from top to bottom). Rooms look best that way, and you'll also have an easier time making aprons, since you can cut all of them from the same board. When you calculate the length of a given apron, allow for the fact that you may want the side casings to be inset $\frac{1}{8}''$ or so from the inside edges of the window jambs. (Look at the diagram above.) If you have to put padding under the apron, cut the padding now.

You can tack it either to the apron or to the wall as a convenience, or you can simply nail through the apron and padding at the same time. Just make sure the edges of the padding are flush with those of the apron, so that together they look like a single piece of wood. Also make sure that the top of the apron is flush with the uppermost edge of the sill. You'll nail the stool across the apron *and* the sill, and it's important that the stool be level.

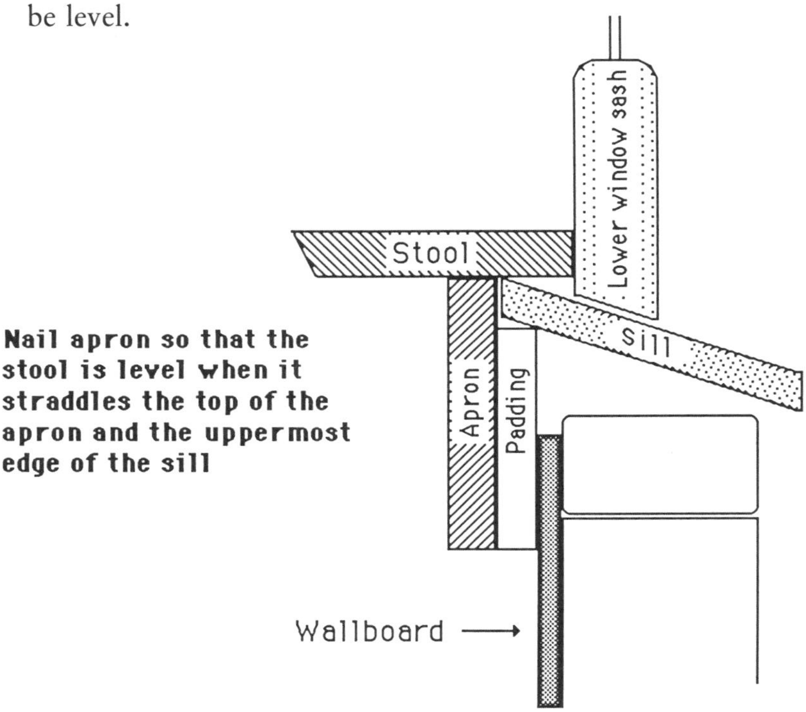

Nail apron so that the stool is level when it straddles the top of the apron and the uppermost edge of the sill

Window sill

Stool

Lower window sash

Sill

Apron

Padding

Wallboard →

Use finishing nails long enough to go through the apron, padding, and wallboard and still bite about 1" or so into the framing around the window. Six nails ought to suffice for each stool—one in each of the corners and another couple somewhere in the middle.

Incidentally, although the apron shown is a rectangle, you might want to cut yours differently. You can round off the two lower corners or make who-knows-what other ornamental cuts. You can rout patterns into the aprons or whittle bas-reliefs of cherubs, ducks, or deities. It's seldom that anyone gets the chance to create built-in room ornaments from scratch, and, if time permits, you might want to use the chance.

Window Stools

With the aprons in, you can proceed to cut the stools. Make them wide enough to accommodate a coffee mug but not so wide that they project too conspicuously into the room. Generally stools are cut a little longer than the aprons underneath, so that they extend about 1" or so to the left and right of the side casings. A stool butts right up against the lower sash of a closed window (the sash is the wood—or metal—that actually encases the glass), so this is the basic pattern you'll have to cut in order to make one . . .

Simple stool pattern (leaves gap between wallboard and ends of stool if jambs aren't flush with wallboard)

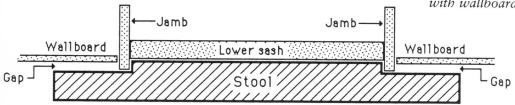

As the sketch indicates, the problem with the basic pattern is that unless the edges of your jambs just happen to be flush with the wallboard, you'll wind up with a little gap behind either corner of the stool. The gaps may not bother you particularly, and they're only visible from certain angles. If you can tolerate them (I did), it makes cutting the stools very easy. But if it's unalloyed perfection you need, you might want to cut stool patterns that are slightly more complex.

The fancier patterns bring the back edges of the stools flush against the wallboard, a condition the epicure in you may insist upon. The first, simpler pattern is easily cut with a backsaw (the kind that's packaged with inexpensive mitre boxes in hardware stores), but for the more complex patterns you'll have to use something like a saber saw, a jigsaw, or a coping saw. None of those is famous for making straight cuts, but this is one case where the cuts needn't be so flawless. Soon enough they'll be

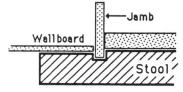

1. If padding isn't yet in place

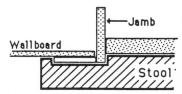

2. If padding is already there

Fancier patterns

concealed on the top by the lower ends of the side casings and on the bottom by the apron.

As with the aprons, you'll get stylistic consistency by slicing all the raw stools off the same board (or similar boards) and then cutting out the notches for each one. If you have access to a table saw or you're good at using a circular saw with the blade tilted a little (virtually all circular saws have an adjustment that lets you tilt the blade), you can cut the three outside edges of the stools at a little bit of an angle. It's more trouble that way, but the result is pleasing to the eye. You can also cut them square and then round the corners with a sander. The nice thing about pine is that being on the soft side, it succumbs easily to touches that warm up its appearance.

Once you've made a stool, you can nail it down with finishing nails. Six nails should do it, eight at the most. (If it feels solid after six, leave it alone.) Half the nails should go down into the apron, the other half into the sill. And before you drive the nails down, make sure the stool edge that faces the outdoors butts smoothly against the lower sash of the window. The window should just barely clear the stool in order to close. To seal against infiltration of air and water, you may want to put a bead of

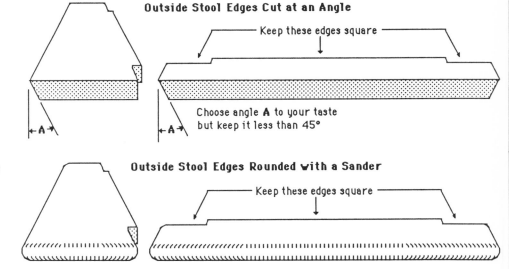

Stool samples

caulking along where the stool meets the windowsill, but you can also caulk after the stool is nailed. You'll also have to nip off a piece of side stop on either side of the window so that the stool can butt all the way to the lower sash. The side stop is the inside rail of the track along which the window sash travels. Take off just enough at the bottom so that the stool can meet the sash. And when you put nails into either end of the stool, stop for a moment to imagine where the lower ends of the side casings will go. If you can drive the nails so that their heads will be concealed by the casings, so much the better.

Side and Top Casings
With the stools in place, the complicated part of the trimming is done. Now you can cut the side casings to the right length, offset them slightly from the jambs if that's what you've elected to do, and nail them on. The right length, of course, is the length that causes the offset (if any) along the side jambs to be matched by the offset (if any) along the top. You can make either miter joints or butt joints for the top corners of the casing.

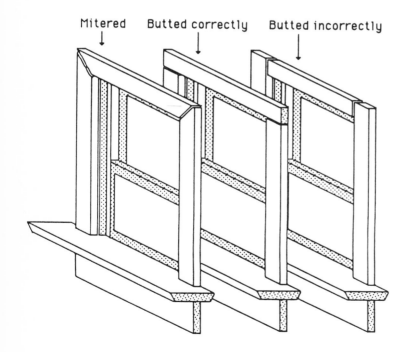

Window casings

Butt joints are best for plain (rectangular-cross-section) trim, miter joints for trim that's milled with curves or angles. If you use butt joints, though, take care to have the top casing sit across the upper ends of the two pieces of side casing. (See diagram on previous page.) Doing it the other way around looks funny.

When you nail the casing pieces to the window, pad them first if you have to and then put the nails in in pairs, sinking one into the jamb and the other into the framing around the window. A pair at either end of each piece and another pair every 2' or so should hold the casings solidly. As with trim in general, finishing nails are the nails of preference.

Trimming the Door

So much for the windows. The door's a lot easier, since all it wants is casing around the sides and top. Remember not to bring the side casings all the way down to the subfloor. Normally one installs flooring *under* the side casings of a door, not around them. So leave a gap between the bottoms of the side casings and the subfloor that equals the thickness of the flooring you plan to put in. And if no such plan has taken shape as yet, this is a good time to read ahead and make one.

Trimming Under the Ridge

We spoke earlier about a piece of trim to cover the gap under the ridge of the ceiling. Clearly the trim runs the full length of

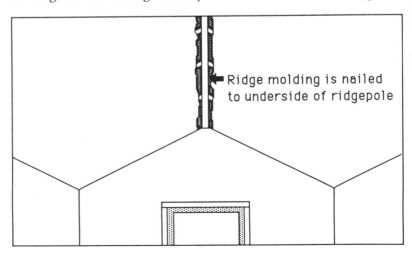
Ridge molding is nailed to underside of ridgepole

the room, and while it looks best if you make it out of one continuous board, that isn't always possible. If you see something appropriate at the lumberyard, by all means grab it. But you can also rip it yourself from a plain pine board. The edges should butt tightly to the wallboard, and I'll leave the choice of nail intervals to you. As the sketch on the previous page suggests, this is one place where stock with a rectangular cross section is least desirable, but even if you're ripping your own trim you can set your sawblade at an angle and then do some further shaping with a sander. You'll find that if you cut the ridge trim ever so slightly longer than the room, it will butt into both short walls with a satisfying snap when you nail it in. Just bend it so the edges fall into place and then let the nails pull it straight.

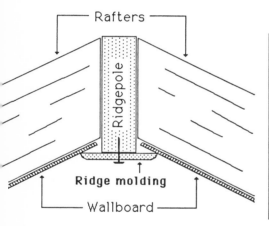
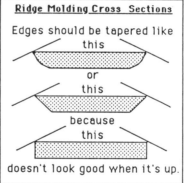

Ridge molding cross sections

Baseboards

The last pieces of trim you have to make (as distinct from purely ornamental pieces that you *choose* to make) are the baseboards, so look around at some existing baseboards, see what appeals to you, and pick a width that's at once pleasing to your eye and sufficient to cover the bottoms of the wallboard sheets. The two most popular kinds are "square" (not really square; the cross section is a rectangle) and clamshell. Lacking a flat upper surface, clamshell doesn't collect as much dust as square molding does, but, aesthetically, I'm not one of its most enthusiastic fans. There's also the fact that you can make square molding yourself, whereas you have to go out and buy clamshell. Even so, those are but two of a great many available baseboard styles, and I suggest you browse the lumberyard before committing.

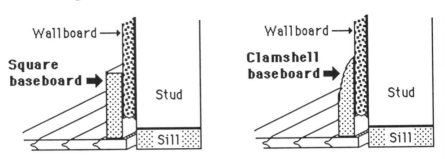

When you cut baseboard, try to use full lengths wherever possible, and where lengths exceed 5' or so cut it a few hairs on the long side. It's inclined to shrink lengthwise as it dries indoors, and you can always get longish pieces to fit by bending them and then nailing in the bellies. Where using full lengths isn't practical, join two pieces with a 45° bevel. And measure lengths of baseboard on the assumption that the joints in all the corners will be butts. If you're using baseboard that's milled in any other way than square, you'll want to cope one of the butt ends in each corner with a coping saw. Coping gives a neater, tighter fit than mitering. At least it's tighter if you remember to cut the baseboard so it's the right length *after* you've coped it.

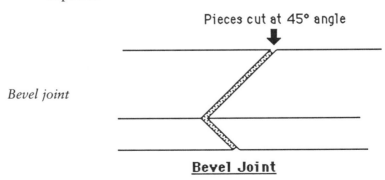

Bevel joint

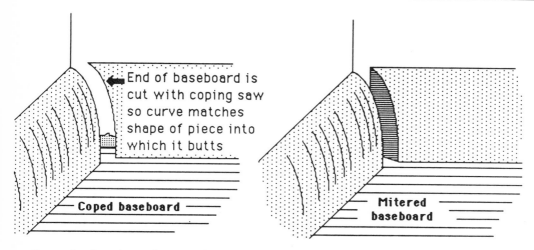

Once the baseboards are all cut, try them out but don't nail them down yet. That part doesn't come until you've installed the flooring, and the best time for flooring is right now.

FLOORING

The three kinds of wood flooring we'll discuss are plywood, pine, and oak strip. Plywood is the most versatile (if not the most attractive), since you can use it as underlayment for vinyl tile, ceramic tile, linoleum, or carpeting, or you can paint it with floor paint and then just walk on it as is. If you want it as underlayment for carpeting or ceramic tile it needn't have a "good" side. In fact, it needn't even be plywood. Chipboard or particle board are usable as underlayment too. Otherwise the upper side should be at least grade B. (Vinyl laid on plywood will collapse into knotholes.) One factor that may play a role here is price. If it's pine or oak strip floors you crave, check to see how much the different grades will cost. Sanded and then varnished or polyurethaned they can look very good, but the volatile prices of even the coarsest grades of pine and oak flooring may tilt you back toward the simplicity of painted plywood. (Yes, a few years from now you can put pine or oak flooring on top of plywood, but I'm giving odds that you'll never get around to it.)

Plywood Floors

Plywood floors have the additional advantage of requiring a minimum of labor to install. You buy the plywood (at least $\frac{1}{2}''$ but $\frac{3}{4}''$ is solider and $\frac{5}{8}''$ is a decent compromise), sweep the subfloor, put down a layer of 15-pound builder's felt, cut the sheets, set them down so the grain crosses the grain of the subfloor and the seams don't fall right over the subfloor seams, butt them as tightly as you can (tongue-and-groove plywood is good for flooring), then nail them with flooring or Sheetrock nails, trying to get as many nails as you can into the joists below. Work toward keeping the hairy, non-factory-cut edges at the perimeter of the floor, because the perimeter will soon be covered by the baseboard. It all takes hardly any time at all, and when you're done you can retire (maybe) from your brief career in carpentry.

Pine and Oak Strip Flooring

If you're laying oak strip or pine board flooring, the process isn't really that much more complex. Tack on a layer of builder's felt and then lay the floor *across* the joists, making sure the starter course is absolutely straight. Both pine and oak strip flooring have tongue-and-groove edges, and when you lay the starter course, keep the tongue side facing toward you. Take care that the back edge of the starter course will wind up under the baseboard, and face-nail the strips so that the nailheads *also* fall under the baseboard. Still more things that should fall under the baseboard (this time the baseboard at the *sides* of the floor) are the ends of all the courses. Once the starter course is face-nailed, toenail into the tops of the tongues. The nails shown in the diagram at top of facing page are finishing nails, which are okay for pine but not for oak. Oak strip is nailed with nails made especially for oak strip; your lumberyard or hardware store will likely carry them. Wherever possible, sink your nails not just through the subfloor but down into the joists. It's helpful to mark the centers of the joists on the felt with a chalkline before you start hammering.

Eventually, course by course, you'll bring the flooring from one edge of the room to the other. To minimize gaps between

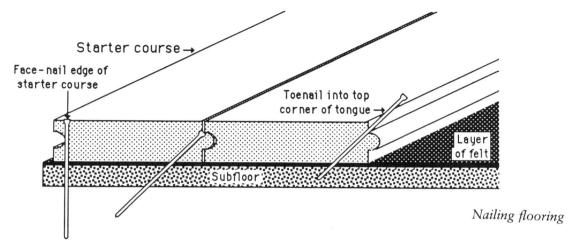

Nailing flooring

the courses, use tongue-and-groove scrap and a hammer to bang each course into the one before, but don't hit the flooring so hard that you smash the tongues. When you're finishing up, make sure the flooring comes close enough to a wall so that the final edge will be concealed by baseboard, and where you have to bring the flooring to a doorsill, bend and then cut the flashing that you long ago installed under the door so that its edge is nearly flush with the surface of the flooring. The lip you'll create will prevent any stray water from trickling under the door and into the crease between floor and subfloor.

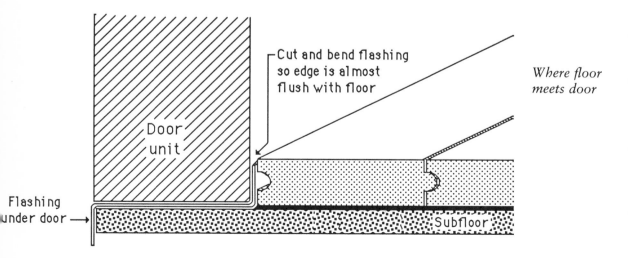

Where floor meets door

With the flooring in place—whatever it happens to be—you can go ahead and nail the baseboards. As for the installation of carpeting, vinyl tile, ceramic tile, or linoleum, seek advice from those who profess expertise (even if, as a rule, the expertise you get is generously laced with folklore). Now that you've trimmed and floored a room, you've got sufficient credentials to propagate some folklore of your own. And it's a cinch that you'll be asked to do so. I have—many times.

16

Paint, Heat, Finishing Touches

PAINT

Here and there across the world, miles from the nearest outbuilding, there are pockets of civilization where paint is regarded mainly as a decorating tool, a convenient way of using color to create subtleties of mood and light and texture. But venture out beyond those prim salons, find some building site that's near completion, pick your way through the rubble of tools, wood scraps, bent nails, and ragged shreds of packaging piled knee-deep around some virgin edifice, and you may very well see paint as something else entirely. Presented with all that bare wood, maybe you'll see it secondarily as a preservative. But first and foremost you'll want to use it as the Ultimate Eraser, that which can bring such disparate things as trim, caulk, siding, filled knotholes, and you-name-what-else together into smooth planes and sharp corners. In the chapter on making a sketch we spoke about drawing with lumber, using the wood itself to sketch the outbuilding. Now that you've sketched one, you'll undoubtedly want to clean up the drawing.

It used to be that when people needed a paint job they almost always hired a painter. Paint was seen as scary and a little baffling; it took professionals to diagnose the need for eye of newt and add it in the right proportions. Since hardly anyone but painters painted, paint stores were relatively scarce, and those that did exist sold mostly to the trade. Today paint is widely advertised in all the media, and the people shown applying it are people "just like you." They've bought their paint in one of those home-improvement supermarkets that sprawl throughout the suburbs or maybe one of the boutique paint stores that you find in cities, and now they're having grand fun rolling it on. Above all, the ads say that painting is easy; anyone can do it. Well, for the past few months, as a novice putting up a building, you've been anything but anyone, and being anyone may come as a relief. So may doing something easy. Besides, it's best to paint at least the outside window trim before mold and mildew turn your new wood into grayish fungal sponge.

Exterior Stain

You may even want to attack the siding, in which case that's what ought to come first. To my taste, rough-sawn siding looks better stained than painted and looks *still* better naturally weathered. By stain, incidentally, I'm speaking of *exterior* stain, not the stuff you smear on unpainted furniture. You can get it with an oil or latex base, and, having far more body than interior stain, it presents a much thicker skin to the weather. If that's what you want, the earlier in the life of the outbuilding you put it on the better. Right now the color of the siding is pretty uniform. But as it weathers it will gray unevenly, and the unevenness may show right through the stain. Staining bare T-111 is about the easiest form of housepainting I know of, especially if you use the latex kind. (I'm an advocate of latex paint in general, and if you start with it from scratch you won't encounter the peeling problems that may ensue years down the road if you try putting a latex-based product on top of an aging oil-based coat.)

Wood Preservative

My own plans for the outbuilding are to let it weather to gray (some sides of the building will get there before others) and then coat it with a clear wood preservative, of which the market offers many different brands. The outside trim is white cedar, and I like the silver-gray it weathers to. The T-111 should weather to a darker gray, which, if all goes well, will make for contrast. If I were to apply a clear preservative right now, the wood would weather anyway but not as fast. Clear preservative is supposed to protect against rot, not the color change you get from sheer exposure. In fact, one school of thought holds that wood on which a clear preservative is merely painted (as opposed to dipped and saturated) won't last much longer than if you leave it bare. Maybe so. But rough-sawn wood, whose many wood hairs give it an enormous surface area, can wick up a lot of preservative. And I like believing that wood preservative works. Placebos can be strong medicine if you believe in them.

Windows

If you've used new door or window units, the outside trim and sash are probably made of smooth-sawn pine (unless you've sprung for bigger bucks and got the kind that's vinyl-clad), which wants some paint (a rubric that includes clear marine varnish or varnish stain). So touch up any caulking around the perimeters of the units, and then, to make the trim look good, sink the outside nailheads with a nail punch. (If you haven't got one, the point of a 16d nail will do nearly as well.) Fill the holes with putty or, even better, with one of the new plastic hole fillers that are showing up in hardware stores. Then put a coat of exterior primer on the bare wood—unless they're factory primed already—and follow it with a coat of exterior trim paint. (Again, my vote goes to latex in both cases.) Of the various surfaces that paints will give you (flat, satin, gloss, etc.), I favor semigloss for trim. It sheds dirt and water better than flat but doesn't glare at you. Unless the primer is of a very different color than the trim paint (not a good practice anyway), one coat of each should be quite sufficient.

Trim

If any of the rest of your exterior trim (fascia, soffits, corner boards, etc.) isn't made of cedar or redwood or one of the other self-preserving woods (teak is most durable of all), you're best off painting it too. Otherwise you retain the option of painting it, but you can also stain it, coat it with a clear preservative, or simply leave it alone. To me, painted rough-sawn cedar looks odd. The protruding wood hairs make it resemble the back of a human hand that's just been dipped in paint. Stain looks more presentable to me, but naturally weathered rough-sawn cedar is still my trim of preference.

Weather Considerations

And a caveat that applies to any outdoor painting: don't do it when it's cold or when there's any likelihood of rain. The effects of rain are obvious enough on water-based latex paint, especially if you're painting only trim and want to leave the siding clear. But rain can damage fresh oil-based paint too, especially if moisture gets under it before it's thoroughly dry, so don't play chicken with the weather forecast. As for cold—and that goes for above-freezing cold, not just frost (refer to the paint can for the allowable temperatures)—it can seriously reduce a paint's ability to stick to wood. Even if you have to hold off painting for a whole season—maybe even two seasons—wait!

Interior Walls

For inside walls the cheapest latex primer is fine as long as you don't also demand that it serve as vapor barrier. If you want it to be vapor barrier too, spend the few extra dollars for a primer that's rated accordingly. Either way, match the color of the primer as closely as you can to the paint you plan to put on top of it. For the final coat I recommend latex still again, but there's latex and latex. The better latex paints can easily cover similarly colored primer in one coat, but some of the cheaper ones can't. This is one instance where money spent can lead to time and money saved on a second coat.

If you're painting the interior trim to match the walls, you'll probably want to use flat paint on the walls and interior semigloss on the trim. So think ahead and buy a brand of wall paint that also offers semigloss in the same color. You'll maximize the chance of getting a good match. And you may not want to prime the inside trim with latex primer. Water-based primers tend to raise the wood grain and give inferior protection against bleeding knotholes. While the perfect knothole sealer hasn't been invented yet, the interior wood primer I like best is shellac. It goes on easily, dries very fast, and seals wood smoothly and effectively. It's carried in an alcohol solvent, so make sure you keep the windows open if you use it. And be aware that shellac can be labeled either as shellac or primer. If it's sold as shellac, don't expect white shellac to be white; it's more or less colorless. To get it with pigment mixed in—a good idea if you're using it as primer—buy the kind labeled specifically as primer. Your paint purveyor will explain further. Then paint over it with whatever paint you plan to use. Interior latex semigloss would be my first choice.

Clear Coats for Interior Trim

Of course, if you have nice pine trim and prefer that it look like what it is, you can seal it with interior wood stain or with a clear sealer, which is much the same substance as wood stain but without the pigment. Then you can put on one of the many oil finishes that are made for furniture (the ones with names like Olde Yorkshire Antique Oil). Or you can skip the sealer and use at least two coats of oil, since any decent oil finish is a sealer, too. Or you can seal the trim with sealer and then put on a coat of paste wax. Or you can seal it with shellac and then wax over the shellac or even *not* wax over the shellac. Or, to get a really long-lived finish and lots of heightened contrast in the grain, you can put on a couple of coats of clear or tinted polyurethane, making sure to thin the first coat according to the manufacturer's directions. Which is all by way of saying that you can treat interior trim just as you'd treat unfinished furniture.

Coating Floors

As for the floor, in the last chapter we touched briefly on paint and polyurethane. If you paint you can use either latex or oil-base, but try to use paint sold specifically as floor paint. It's tougher than the other kinds and can tolerate more traffic. If you want polyurethane (or even varnish, which seems to have gone a little out of fashion but still gives a nice finish), you'll have to rent a floor sander to clean and smooth the wood. Sanding oak can be slow because the wood is hard, but apart from that it's fairly easy. Pine poses more of a challenge because it's so soft that floor sanders cut grooves in it. You can sand the grooves off with the same sander that cut them in, but if you aren't careful you can wind up grinding half the thickness of the wood away. Sanding pine with a floor sander is one process in which experience is a tremendous help, so, if that's what you intend to do, plan on making small amounts of experience go a very long way.

Otherwise, floors are inherently easier to coat than walls. Being horizontal, pools of paint or polyurethane or varnish are automatically self-leveling. Still, don't let that carry you away. There's asphalt-saturated felt just under the flooring, and virtually any nonaqueous solvents in the coating you apply are capable of seeking out the underlying asphalt and wicking it up to the top of the floor, where the black streaks it leaves will soak in and stay forever. Coat the floor, but don't flood it!

HEAT

Short of running a duct between your house and the outbuilding, you'll probably want to install some sort of heat source. If it's only occasional heat you need—let's call it touch-up heat—an ordinary electric floor heater may be just the thing. All you do is buy it, plug it in, and wait for the room to warm up. It's even likely to come with a built-in thermostat, and if you've insulated the building you'll be surprised at how effectively it heats. Of course, you may also be surprised at how much electricity it gobbles up. A typical electric heater will draw in the neighborhood of 13 amps, and if you've put the outbuilding on a 20-amp breaker that leaves you only 7 amps for other uses before you trip the breaker. Still, 7 amps is probably more than enough to run your lights and maybe a radio (a 100-watt bulb draws a bit less than 0.9 amp at 115 volts; the formula is amps = watts/volts), and in the end your principal objection to electric heat, clean and installation-free as it is, will probably be cost. Unless you live in a place where hydroelectric power keeps rates exceptionally low, it could run you hundreds of dollars to heat your outbuilding with electricity through a whole winter.

Kerosene Heat

That unpleasant fact may turn your thoughts to kerosene, which is cheaper, a tad smellier, a bit more trouble, a bit more hazardous, a bit more difficult to regulate, and also installation-free. There are modern kerosene heaters around that can easily heat a building like yours (assuming you've kept it within the dimensions we've discussed). In most parts of the country kerosene is a good deal cheaper to heat with than electricity, and, to save the money, all you have to do from time to time is fuss with the wick and lug kerosene.

Gas Heat

Then there's bottled gas. I like gas heaters a lot: they're clean, fast, self-regulating, and cheaper by far to operate than their electric counterparts, and if your local bottled-gas company puts in a good-sized tank you can go for long periods without a refill. But now we're crossing into new terrain. A gas heater needs to be installed. At the very least you have to vent it and run tubing from the tank to the heater. You (or your gas deliverer) also have to be able to refill the tank or replace empty tanks with full ones, and if you've built your outbuilding deep in the wilderness that may disqualify it from consideration. But if it's steady, winter-long heat you want and gas is practical, consider it. Modern heaters come in far too many shapes and sizes for me to tell you how to install one: some go through an outside wall (which means you'll have to carve up one of your walls), some freestanding ones go right into the room, and some even go under the building and send up heat through a floor register. Bottled-gas companies have catalogues, and other retailers sell gas heaters too. You can figure on installing a heater yourself and leaving the hookup and the installation of the tank to the gas company. The heater can be expected to come with instructions that anticipate almost any kind of installation.

Wood Stove

Finally, for the primally self-reliant among you, there's wood, which, with its multitude of drawbacks, still remains appealing. If you need consistent, season-long heat and have to buy firewood at the prices that dealers charge vacationers and weekend residents for "fireplace logs," dismiss it out of hand. At those prices you may as well use gas or kerosene or even electricity and spare yourself a lot of trouble. If your neighbors are close enough to mind the smoke, or you're repelled by the idea of scouring creosote out of stovepipes, or you hate having your floor constantly littered with branchlets and bark chips and wood grubs both dead and alive, or you're not prepared to countenance incessant interruptions of whatever you'll be doing in the outbuilding because you have to go get wood or cut wood or stack wood or buy wood, then you're probably too civilized for wood and would be best off with something else. But if the air around you isn't already choked with woodsmoke and you've got a productive woodlot, or you can buy wood at decent prices, or your outbuilding is surrounded by deadfall that practically insists on being converted into heat, or you enjoy the sheer sport of cutting, splitting, hauling, and stacking logs, or you absolutely need to get your heat for "nothing" (it is to laugh), then prepare to spend a lot of money—money that the business world coyly earmarks as "initial startup cost"—and buy an airtight wood stove, some heavy-gauge smokepipe whose diameter matches the collar on the stove, a chimney kit that matches the type and pitch of your ceiling and roof (thimble, chimney mount, etc., about which more later), at least a 3' length of triple-wall or Metalbestos chimney whose inside diameter matches that of the stovepipe, a combination chimney cap and spark arrester to match the chimney, and assorted heat shields, bricks, stone slabs, and whatever else it takes to keep the stove from setting fire to the building.

Heat Shields

It should be apparent from the foregoing list that we can't go through the installation of the stove in any great detail. For one thing, different manufacturers specify different minimum clearances between various parts of their stoves and the nearest flammable surface. (Sheetrock is considered a flammable surface.) But the heat shields that many of those same manufacturers offer (at extra cost) for the back, sides, and bottom of the stove reduce the clearances by quite a lot, so you may want to add some heat shields to your shopping list. You can also cut down on clearances by making heat shields of your own for the walls immediately nearby. A common way to do it is with pieces of sheet metal that are held 1" or more away from the walls by ceramic spacers. You'll also have to separate the stove from any wooden floor, which may mean you'll have to make a hearth by setting a slab of slate or bluestone on bricks and then putting the stove on top of that. Of all places, your local fire department can be a good source of guidance. Many offer booklets that provide clear guidelines on how to install and use a wood stove, including things like the minimum gauge of sheet metal you can use for homemade heat shields and minimum stone thickness for a homemade hearth. Someone else you might consult is your building inspector. Since it may be part of his job to inspect wood stoves in the first place, he can often tell you what you have to do to make yours pass. The best source of all, though, is the literature that accompanies new stoves. Manufacturers are usually pretty careful to provide guidelines that will satisfy any of the uniform fire codes, especially since fires associated with wood stoves have begun to eat into their product's popularity.

Given the clearances they need, any wood stove you buy will also eat considerably into the usable floor space of your outbuilding, so know what you're getting into before you install one. One thing you'll certainly be getting into—all over again—is construction carpentry. Because in order to install the chimney thimble you'll have to cut a hole right through your ceiling and on into the roof.

Chimney thimble

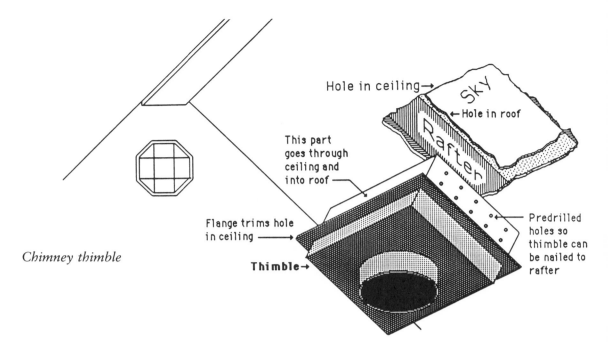

And then you'll have to set a chimney mount over the top of the hole more or less like this:

Chimney mount for roof

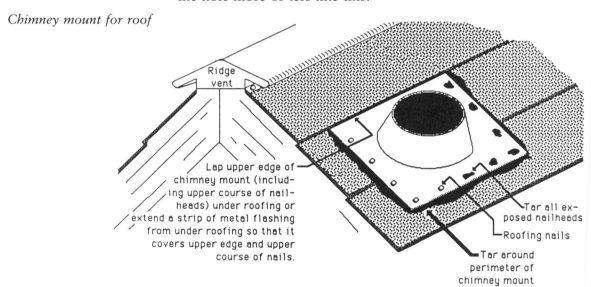

Caution: Chimney mount is shown close to edge of building only for purposes of illustration. In real life a chimney mount that close to the edge wouldn't permit sufficient clearance between the smokepipe and the interior wall. Check with the installation instructions that come with your smokepipe or with your local fire department to determine what minimum clearance can be.

The thimble and chimney mount combine to hold your flue and chimney in place while keeping it away from any nearby wood. I can only guess at what your hardware will actually look like, but what the drawings show is fairly typical. Most thimbles are sized to fit between a pair of rafters on 16″ centers, and their trim flanges are generally big enough to conceal the edges of a crudely cut ceiling hole. When you do buy a thimble (as I said, they often come as part of a chimney kit), verify that it's made for a ceiling pitched the way yours is and that it's of the "smokepipe in, chimney out" variety. Some thimbles are made only for horizontal ceilings, others only for the passage of chimney pipe through a roof. (Just so we're clear on this, smokepipe, sometimes called flue pipe, is made of a single wall of sheet metal; chimney pipe is made either of two concentric walls of sheet metal with insulation in between—that's the Metalbestos type—or three concentric walls of sheet metal with air spaces in between. Both are stunningly expensive for what they are.) In an installation like yours, you run smokepipe from the stove up to the thimble and sit chimney pipe on *top* of the thimble. The thimble contains the union that joins the two.

Notes on Installing a Wood Stove
Now some additional suggestions. Try not to put any elbows in your smokepipe. The more corners that you ask the rising smoke to turn, the less well the stove will draw. And don't run your smokepipe too near a wall. As with stoves themselves, there are minimum clearances between any smokepipes and the nearest flammable surface. You can reduce them with a heat shield made for smokepipes (see the guidelines you've responsibly acquired for what the clearance is), but there's a limit to what heat shields will do. Also, try to keep your chimney fairly close to the roof ridge. The farther down along the roof you cut your hole, the taller the chimney has to be. (Yet again, refer to your guidelines.) If you put your chimney near the ridge, you'll almost certainly be able to get away with making it out of a single 3′ section of chimney pipe. But if you place it farther down the roof you'll probably have to add a second section,

and a 6′ chimney will need guy wires to stand up to the wind, which complicates the installation that much more. If your chimney pipe or cap or spark arrester isn't made of stainless steel (most triple-wall chimney pipe is galvanized), get a can of spray paint made for metal and give them a protective coat. Wood smoke is incredibly corrosive, and you may save yourself a lot of trouble down the road. And when you're stacking sections of stovepipe, do it so the end of each section fits *inside the one below*. Otherwise you'll make it possible for liquid creosote to drip down the inside walls of a given section onto the *outside* walls of the section under it, which is smelly at best and capable of causing fires at worst. For stability, put some sheet-metal screws at all the stovepipe unions, but don't use so many that it's cumbersome to take the flue apart to clean it. And, maybe most important of all, get a decent, newish stove. The cast iron in an old one may be so shot through with metal fatigue that a sudden temperature change may crack it. Besides, wood-stove technology has advanced considerably over the past couple of decades. Get one that takes advantage of the technology.

Before leaving the subject, I'd like to point out that I'm aware of the extent to which everything I've said about wood stoves—the trouble of installing them properly and the additional trouble of using and maintaining them—may steer you toward another heat source even if it's wood heat that you wanted all along. In fairness I should say that I installed a smallish wood stove in the outbuilding you've been seeing pictures of, that for six or seven days a week during the entire winter it kept the building warm enough to let me leave some windows open, that burning deadfall kept my fuel bills to zero, that when I have to go outside on cold, rainy days to bring in wood I wish I had a gas heater, and that I finally prefer the wood stove because, for all its inconvenience, it gives me as much heat as I want and leaves me beholden to no one.

A WORD ABOUT PLUMBING

Throughout this book we've tiptoed around the subject of plumbing, and now we'll skirt it once and for all. A great many townships won't let you put a sink or toilet in an outbuilding anyway, at least not without first getting it zoned as a separate residence. Moreover, plumbing in its infinite variety deserves a book all by itself. If the building were more complex than the one at hand, I might have paused at some intermediate point to remind you that if you *are* going to install plumbing, this would be an appropriate time to drill your studs and put your pipes in the walls. But the outbuilding's walls are all outside walls, and good plumbers don't put pipes in outside walls. Anyway, the simplicity of the floor makes it easy to bring pipes right up through the bottom of the building. Yes, the pipes can freeze and maybe burst, but you can bury feeder pipes below the frost line and wrap them with electric heater tape where they enter the building. The easiest solution may be to run a length of good-quality hose *over*ground from your house and screw a shutoff valve onto the end. The hose will freeze in winter, but for part of the year you'll have running water in its simplest form. You can also carry water jugs. It's been done before and has been shown to work.

LAST WORDS

Finishing a substantial project can be a poignant time. In your rush to completion you've probably had to trade off some familiar old compulsions for new ones, which may mean that you've fallen behind on things like reading mail, paying bills, filing tax returns, keeping up your magazine subscriptions, talking to your family, servicing your car, maybe even shaving, and confronting all of those at once can be poignant in the extreme. In another vein, the new persona that the outbuilding all but forced you to adopt—the competent, inventive, self-reliant designer-builder-artisan that you started to be from the moment you first broke ground—will have to give way in some measure to the old one, to the one you normally employ to live your life. Before that melancholy transition takes place, however, I suggest you finish screwing cover plates onto all the open switches, sockets, and junction boxes in your walls, that you stick louver caps back into your soffit vents if you removed them to paint or stain the soffits, and that, whether you get to it or not, you at least think about closing the space under the outbuilding by nailing some pressure-treated lattice between the building and the ground. The building is still up on stubby stilts, and lattice will bring it down to earth, much as finishing the building may bring *you* down to earth. Welcome back, then, to reality.

Reentry may come with a disorienting jolt. We've likened building the outbuilding to the private pursuits of people like Jefferson and Franklin, and even if the analogy's been stretched a little, it may make your return trip seem like a time-warp crossing from colonial America to the dreaded here and now. One day, confidently swinging your hammer, you feel that you might easily be living in a time when nearly everyone could build a building, when nearly everyone who happened by was pretty much the same sort of competent generalist that you fancy you've become yourself. Franklin wants a wood stove that's also a fireplace? Why not? He designs and builds one, and the idea of it survives for centuries as the Franklin stove. Jefferson wants a baronial manor? Good enough. Serving as his own architect, he designs and builds one, and it survives for centuries as Monticello. A disparate collection of inexperienced politicians want a constitution? Fine. They design and build one, and it shakes the world, becomes the model for a host of others, and survives for centuries as the armature of a government that still aspires to be humane.

One day you're there, and the next day—well—you're here. Of course, the "here" we're speaking of is no place that the founding fathers hoped to see, nor is the welcome it extends to designer-builder-artisans a particularly warm one. "Here" is a world that's corporate to the bone and getting more so, that's breathtakingly unself-critical, and that encourages the development of skills so specialized, so "professional," that one hesitates to call them skills at all. By any reasonable standard, skills should be a source of confidence, but what too many of today's proficiencies induce is merely arrogance. Those amateur colonials, competent at their haphazardly assorted skills, could frequently be confident leaders. Today's professionals are mostly nervy ones. What passes nowadays for competence is often just the ability to steer a self-enhancing path through the hierarchies and bureaucracies of companies, governments, and the social networks that surround them. Real skills take a distant second place to being wised up.

Whatever else it was, making the outbuilding has been an exercise in personal competence; it's been quintessentially uncorporate. If the central concept of the corporation is the notion of limited liability, you, by contrast, have been liable all the way—for the work, the bills, the building's structural integrity, for its look and feel and how effectively it does the job of sheltering. If you've screwed up, there isn't any corporate facade to hide behind. The responsibilities begin and end with you. You can always try hiding behind this book (that's one of the things it's for), but one slim volume won't give you the kind of concealment that a squad of Wall Street lawyers will.

What you get in return for assuming all that responsibility is the dissolution of the mystery inside the walls. Unlike your television set or VCR or your computer or even the hydrogen bomb, the outbuilding won't be just another black box that plays a major role in your life even though its innards baffle you. Your liability may not be limited, but neither is your freedom to chop into what you've built and knowledgeably change it. Your ownership of it isn't mitigated by the need to leave it in the hands of specialists when it doesn't work the way you want it to. It's yours in the fullest sense.

Thinking about Buber's assertion that the purpose of imagination is to imagine the real, it's tempting to take the easy way out and say that the "real" he asks us to imagine has grown complex beyond imagination. I suspect that isn't so. Our imaginations are more powerful than that. Maybe we have to stop wasting them on twaddle and use them more responsibly. I hope the outbuilding isn't merely twaddle, that in its modest way it's more of a beginning.

Glossary

APRON. In trimming the insides of windows, normally the lower most piece of trim. It runs across the bottom of the window just underneath the stool and faces into the room. (*See also* STOOL, CASING.)

BACKSAW. Handsaw normally used in conjunction with a miter box. It rides in the grooves of the miter box and thereby cuts trim, molding, etc., at a fixed angle.

BASEBOARD. Molding or plain board that runs along the bottom edge of an interior wall. It trims the joint between the wall and the flooring just beneath it.

BATT. A rectangularly shaped piece of fiberglass insulation. Its length may be 4′ or greater, and its width is normally somewhat less than 16″, so it can fit inside a typical bay of a wall or ceiling. (*See* BAY.)

BAY. The rectangular space between two adjacent studs in a stud wall, between two adjacent rafters in a roof, or between two adjacent joists in a floor. Bays are most commonly $14\frac{1}{2}″$ wide.

BEAD (of caulk). *See* CAULKING.

BEAM. A load-bearing horizontal structural member. (*See also* POST AND BEAM.)

BIRD'S-MOUTH. A notch cut into a rafter near its lower end so that the rafter can properly seat onto the wall that supports it.

BOXING. Boards nailed over the ends of the joists of a floor. When their ends are covered in this fashion, the joists are said to be "boxed in."

BUTT, BUTTED, BUTTING. Surface-to-surface contact between two boards. There are many ways in which boards can be joined together, some involving the creation of elaborate grooves and tracks within the surfaces to be joined. The simplest way is with an ordinary butt joint. Butt joints use only the surfaces that are already there.

CANTILEVER. Any rigid construction that extends horizontally beyond its vertical support.

CASING. In trimming windows, whether inside or outside a structure, collectively the top and side pieces of window trim. The top horizontal piece is the "top casing," and the two vertical pieces along the sides are the "side casing."

CAULKING. Material used to close seams and seal cracks in a structure. Most modern caulking materials have a latex base; some have silicones added for greater durability. Where caulking will be exposed to the elements, an appropriate exterior grade should be used. A uniform strand of caulking material is called a "bead" of caulk.

CENTERS. Studs or joists or rafters "on 16" centers" are spaced so that imaginary lines drawn longitudinally through the centers of the boards will be exactly 16" apart.

CLAPBOARD. Siding made of overlapping horizontal boards. Clapboards are usually milled thicker along one edge than the other, and the thick edge of each board generally laps over the thin edge of the board below.

COPE, COPED. Cutting boards according to a curved pattern. Coping is often used to join boards, especially at an inside corner, that have been milled to have a curvilinear cross section. The end of one of the boards is coped to fit the curved cross section of the other. The cut is generally made with a coping saw.

CRIPPLE. A vertical member of a stud wall that is cut from the same lumber (normally 2x4) as a full-length stud but which doesn't run the full height of the wall. The sections of 2x4 that run vertically between the bottom of the rough opening of a window and the bottom of the wall itself are cripples.

CROWN, CROWNED. Natural, largely unavoidable, curvature in milled lumber, often the result of uneven drying.

DIMPLE. In gypsum wallboard, a shallow depression left around a nail by the head of a hammer. The dimple is normally filled with wallboard compound, thereby concealing the nailhead.

EAVES. The overhanging lower edge of a roof.

FASCIA. Exterior trim along the lower edge of a roof that closes part of the ends of the rafters. Its broad side is vertical and therefore faces outward. This distinguishes it from soffits, broad sides of which are horizontal and therefore face downward. (See SOFFIT.)

FLASHING. Metal or plastic strips that cover or otherwise protect exterior joints between two or more wood surfaces or between a wood surface and one of stone or concrete. By keeping such joints dry, flashing discourages leaks and inhibits rot.

FOOTING. Any flat piece of stone or concrete, usually poured or set in bare ground below grade, on which a foundation or load-bearing post can rest securely.

GABLE. That part of the wall of a building under the end of a roof with two sloped sides meeting at a central ridge.

GFI. *See* GROUND FAULT INTERRUPTER.

GIRDER. A major load-bearing horizontal structural member, and therefore also a beam. In a frame house, the floor joists are usually supported by at least one girder.

"GREENIES." Special green wire nuts with an "out" hole on top that allow you to connect several bare ground wires to a single ground screw. You splice a new length of bare wire to the other ground wires, then slip a greenie over it and twist it onto the splice so that the new bare wire comes out through the "out" hole in the greenie. The single length of bare wire is fastened to the ground screw, and now all the ground wires in the splice are grounded.

GROUND FAULT INTERRUPTER. Special circuit breaker built into electrical outlets for use where moisture may be present (outdoors, near bathroom sinks, etc.). GFI receptacles are installed in the same manner as conventional outlet fixtures but give additional protection against accidental shock. Like ordinary circuit breakers, GFI's trip at a given number of amps.

HEADER. A horizontal, load-bearing structural member most commonly used across the top of a rough opening for a door or window. (*See* JACK, ROUGH OPENING, TRIMMER.)

JACK. A vertical, load-bearing structural member in a stud wall whose purpose is to support a header. Typically, a header is supported at either end by a jack.

JAMB. Either of the sides of the fixed frame in which a door or window is set.

JOISTS. The basic horizontal, load-bearing structural members in a floor.

ORTHOGONAL. Perpendicular, at right angles to.

OVERHANG. The amount by which the lower edge of a roof extends past the outer wall of a structure.

PITCH. The slope of a roof, normally expressed in inches per foot. A roof pitched at 8" per foot rises 8" vertically for each horizontal foot it traverses.

POST. A major vertical, load-bearing structural member.

POST AND BEAM. A style of construction in which most of the load of a structure is carried by horizontal beams supported by vertical posts.

PRECUTS. 2x4 studs cut to a length of $92\frac{5}{8}''$. When normal stud walls are made with precuts, they give a finished ceiling height of 8'.

R.O. See ROUGH OPENING.

R-VALUE. A measure of the insulating power of various materials. The $3\frac{1}{2}''$ fiberglass insulation used in stud walls made of 2x4s is normally rated at R-11. The larger the R-value, the greater the ability of a material to inhibit radiant heat loss.

RAFTERS. The basic structural members—the "ribs"— of a roof.

RIDGE. The horizontal line along which the tops of the rafters of a roof meet.

RIDGEPOLE. A board running horizontally along the ridge of a roof. The tops of the rafters butt into the ridgepole.

RIDGE VENT. A device for ventilating a roof by allowing air trapped inside the roof to escape through protected openings along the ridge.

RISE. The distance, measured along the vertical, between the lowest point and the highest point of a roof or a flight of stairs. (*See also* RUN.)

"ROCKING." Slang term for installing gypsum wallboard. (*See* SHEETROCK.)

ROMEX-TYPE CABLE. "Romex" is a brand of electrical cable that has become generic. It's the kind of cable normally used for inside-the-wall household wiring.

ROUGH OPENING. (Commonly abbreviated R.O.) An opening in a stud wall, normally rectangular, for installing a door or window.

ROUGH-SAWN. Sawn in such a fashion as to leave an intentionally rough surface. Rough-sawn lumber is often used for exterior trim and siding when the builder plans not to paint it.

RUN. The distance, measured along the horizontal, between the lowest point and the highest point of a roof or a flight of stairs. (*See also* RISE.)

SASH. In a window, the framework in which the panes of glass are set.

SCAB. A short board used to tie two longer boards together at their ends.

SELVEDGE. In regard to roofing materials coated with mineral particles, that part of a roof shingle or a sheet of roll roofing that is not so coated. The selvedge is generally lapped under the coated part of a neighboring sheet or shingle.

SHEATHING. Material nailed over the outside of an exterior wall of a structure. It may consist of boards or sheets of plywood or sheets of some other composition material manufactured for the purpose.

Normally siding is then nailed over the sheathing. (*See* SIDING.)

SHEATHING/SIDING. Sheets, usually of plywood, intended to serve as both sheathing and siding.

SHEETROCK. A brand name of gypsum wallboard that has become generic.

SHIM. A sliver of wood used as a spacer.

SIDING. The outermost material of an exterior wall.

SOFFIT. Exterior trim along the lower edge of a roof that closes part of the ends of the rafters. Its broad side is horizontal and therefore faces downward. This distinguishes it from fascia, the broad sides of which are vertical and therefore face outward. (*See* FASCIA.)

SPACKLE. Brand name of wallboard compound that has become generic.

SPLIT-SHEET ROLL ROOFING. Roofing material that comes in 3'-wide rolls. The sheets are covered with mineral particles along half their width. The mineralized part is exposed to the weather, while the selvedge (*see* SELVEDGE) is lapped under the adjacent course.

STOOL. In trimming windows, the shelf-like horizontal piece near the bottom of the window. Often erroneously called "window sill."

STRINGERS. In a staircase, the cut-out frames to which the treads are nailed.

STUDS. The basic vertical, load-bearing structural members of a wall.

SUBFLOOR. Boards or plywood sheets nailed directly over and onto the joists of a floor. The final flooring is installed over the subfloor.

T-111. The most common kind of sheathing/siding. (*See* SHEATHING/SIDING.)

TEMPLATE RAFTER. The master rafter, cut very carefully, from which the other rafters in a roof are copied.

THIMBLE. An insert in a wall or ceiling, usually of sheet metal, which allows the metal chimney or flue pipe of a stove to pass safely through the wall without danger of fire.

TOENAILING. A technique for nailing boards—usually studs or joists—into place when face-nailing or end-nailing aren't possible. A nail is driven obliquely into the side of a stud near its end. The nail is hammered through the stud, out the end, and then sunk angularly into the board to which the stud is being nailed.

TONGUE-AND-GROOVE, TONGUE-IN-GROOVE. A joint between two boards in which a "tongue" milled into the edge of one board fits into a corresponding groove in the edge of the second board to produce a flush surface.

TRIMMERS. In a rough opening in a stud wall for a window or door, the vertical sections that line the opening. Usually cut from the same size lumber as the studs, they run down the insides of the opening from under the ends of the header to bottom of the opening. (*See* ROUGH OPENING.)

UNDERLAYMENT. Plywood or other composition board nailed over a subfloor when carpeting, floor tiles, or linoleum are to be installed.

WALLBOARD. Sheets of gypsum or plywood that serve as the interior walls of a room. Wallboard is nailed onto the inside surfaces of a stud wall.

Index

A-frames, 109–11
Aggregate, 40, 42, 43
Air conditioning, 158
Alternating current (AC), 160
Aprons, 199–201, 205–7
Asphalt shingles, 135–36
Attic, 20
Awning, 152–54

Banisters, 149
Barns, 14
Basements, 24–25
Baseboards, 186, 190, 202, 211–13, 216
Beams, 20
Bearing walls, 15
Bird's-mouths, 19, 108, 110–11
Boards, *see* Lumber
Boxes, electrical, 167–68
Boxing, 17, 34, 67, 70, 120
    of exterior deck, 144–45, 147–48
    ordering materials for, 56–57
Bridging, 16–17
Buber, Martin, 1, 3, 11, 26, 231
Builder's discounts, 48
Builder's felt, 21, 115, 117, 142

    under flooring, 214
    installing, 122–26
    for roof, 135
Building permits, 25–26, 28, 35

Cantilevering, 32–33
Carpeting, 213, 216
Casement windows, 76
Casings, 199–205, 209–10
Caulking, 126–28
    of roof, 135, 137
Caulking gun, 7
Cedar, 139
Cedar shingles, 135
Ceiling, 158
    height of, 82
    paneling of, 185, 186
    Sheetrock on, 185, 190–92
Ceramic tile, 213, 216
Chimneys, 224–28
Chipboard:
    for sheathing, 115
    for underlayment, 58, 213
Circuit breaker, 174
Circular saw, 7, 200
Claw hammer, 7
Common nails, 54–55

239

Concrete, 39–45, 146
Corner boards, 142, 219
Corner posts, 95–98
Cost, rough estimate of, 31
Cross-bridging, 17

Deck, 10–11, 17, 23, 46–74
    building, 59–74
    exterior, 144–49
    materials for, 48–59
    nailing soleplates to, 92–93
    and size of building, 30
    in sketch, 28, 29, 32–35
Decking, roof, 130–34
Design, 14–21
    of stairs, 151
Direct current (DC), 160
Doors, 27, 75–80, 115
    framing, 87, 89–90, 94
    installing, 117–20
    insulating around, 183
    light switches near, 164
    lumber for openings, 82
    in sketch, 28
    trimming, 202–4, 210
Double-hung windows, 76
Double plates, 17
Douglas fir, 83, 107
Drip edge, 134–35

Eastern-style framing, 30, 98–100
Electric heaters, 222–23
Electrical outlets, 156–57, 160
    heights of, 163–64
    spacing of, 164
Elevations, 35

Fascia, 19, 106, 141–42, 219
Feeder cable, 171–73
Fiberglass insulation, 158, 178
Flashing, 53
    of gable ends, 126–27
    of girders, 65–66
    around windows and doors, 75, 117–18
Flooring, 156, 198, 202, 213–16
    coatings for, 221–22
Floor joists, 15–18, 20
Footings, 39–45, 146
Foundation, 10, 15, 25, 36–45
    clearing brush and trees for, 37
    concrete for, 39–45
    digging postholes for, 38–39
    laying lines for, 36–37
    marking out posts for, 37–38
Frame design, 14–21
Franklin, Benjamin, 188, 230
Freemasons, 188
Furring strips, 135

Gable ends, 80
    flashing and siding, 126–27
    framing, 113
    sheathing/siding for, 117
Gable nails, 113–14
Gas heaters, 79, 223–24
Girders, 15, 17
    cutting, 63
    flashing, 65–66
    leveling and trueing, 63–65
    ordering materials for, 51–53
    and size of building, 30–31
    in sketch, 32–33
    toenailing joists to, 68–69
    yoked to posts, 66
Ground-fault interrupter (GFI), 174
Ground rod, 175
Ground wires, 170

Hammer, 7
Headers, 81–83, 86, 87, 90
    cutting, 89
    installing, 91–92
Heating, 222–28
Heat shields, 225

Heel cut, 19
House sills, 15, 17

Insulation, 21, 57, 156–58, 177–83
  installing, 70–72
  vapor barrier and, 180–81, 183

Jacks, 86, 94–95
Jambs, 78
Jefferson, Thomas, 178, 188, 230
Joists, 15–18, 20, 23
  cutting, 66–67
  for exterior deck, 144–45
  installing, 67–68
  ordering materials for, 55–56
  with scabs, 69–70
  and size of building, 30–31
  in sketch, 33–34
  toenailed to girders, 68–69
Junction box, 164, 170–71

Kerosene heaters, 223
King studs, 87, 90–91

Ladder safety, 131–32
Lally column, 15, 17
Level, 7
Light fixtures
  boxes for, 167
  circuit for, 169
Light switches, 164
  exterior, 174
Line level, 7, 49
Linoleum, 213, 216
Lumber, 22–23, 46, 52
  for deck, 48–59
  purchasing, 47–48
  for roof, 106–7
  for walls, 81–83
  scrap, 59

Materials, 8, 22–24
  cost of, 31
  ordering, 46–48
  scavenging, 14, 29
  for sheathing and siding, 115–17
  for stairs, 150
  for taping and spackling, 193
  See also Lumber
Medieval craft guilds, 188
Modularity, 22–24, 29
Molding, 185, 186

Nail-puller, 7
Nails, 54–55

Oak, 83
Oak strip flooring, 213–16, 222
Outlets, *see* Electrical outlets
Outside lights, 164–65
Overhang, 19, 104, 106

Paneling, 156, 158, 184–87
  vapor barrier and, 181
Particle board underlayment, 58, 213
Painting, 202, 217–22
Phone cable hookup, 175–76
Pine, 139
Pine board flooring, 213–16, 222
Pitch of roof, 18, 80, 102–3
Plastic vapor barrier, 57–58
Plates, 17–20, 84–88
  doubling, 100
  lumber for, 81, 82
Plumbing, 229
Plywood, 23–24
  combinating sheathing/siding, 116–17
  for deck, 58, 72–73
  for flooring, 156, 213, 214
  for roof decking, 130–34
  for sheathing, 115
Polyurethane, 213, 221, 222
Portland cement, 40, 42–43

Post-and-beam design, 14, 15
Postholes, digging, 38–39
Posts
   concrete, 45
   cutting true lengths for, 61–62
   for exterior deck, 144–46
   marking out, 37–38
   ordering materials for, 49–51
   rough-cutting, 60–61
   yoking girders to, 66
Primer, 221
Putty knife, 193

Rafters, 18–20, 103, 104
   cutting, 107–9
   full-size drawing for, 104–6
   installing, 111–12
   nailing Sheetrock to, 189
   nailing tie beams to, 112–13
   ordering lumber for, 106
   and size of building, 30
Ridge molding, 210–11
Ridgepole, 19
   lumber for, 107
   nailing rafters to, 111–12
   preparation of, 108
   raising, 109–11
Ridge vent, 134, 135, 138–39
Roll roofing, 135–38
Romex-type cable, 166
Roof, 8–9, 102–14, 129–43
   in frame design, 18–21
   framing, 107–13
   full-size drawing of, 104–6
   installing, 134–43
   insulation of, 158, 179–80
   ordering lumber for, 106–7
   overhang of, 104
   pitch of, 18, 80, 102–3
   in sketch, 28, 35
   working up on, 130–34
Roof decking, 19
Roofing, 135–38

Rosicrucians, 188
Rough openings, 75, 77–80, 86–87
   constructing, 89–90
   cutting lumber for, 85
   framing, 93–95, 99
   in gable ends, 113
   installing windows and doors in, 117–20
   lumber for, 82
Router, 200
Rustic style, 155–56

Sand, 40, 42, 43
Sandpaper, 198
Saws, 6
   circular, 7, 200
Scabs, 56, 69–70, 92, 107, 108
Scaffolds, 103, 132
Scrap lumber, 59
Sealers, 221
Seat cut, 19
Sheathing, 20–21, 115–17, 204
   lumber for, 106
Sheetrock, 21, 23, 156, 158, 184–85, 187–97, 204
   installing, 189–92
   taping and spackling, 192–97
   vapor barrier and, 181
Shellac, 221
Shims, 58–59
Shingles, 20, 58–59, 135–36
Siding, 21, 115–17
   of gable ends, 126–27
   installing, 120–26
   lumber for, 106
   stain for, 218
Sills, 15, 17, 18, 84–87
   lumber for, 81, 82
Sinkers, 54
Siting, 24–26
Size, determination of, 29–31
Sketch, 27–29, 32–35

size and, 29–31
Skids, 130–31
Smokepipe, 224, 227
Soffits, 19, 139–41, 219
   lumber for, 106, 107
Soffit vents, 104, 140
Soleplates, 17, 85, 86, 90, 92, 98
Spackling, 185, 187, 192–97
Split-sheet roofing, 136
Squaring up, 100–101
Stacking, 17, 86
Stairs, 149–52
Staple gun, 72
Stools, 199–201, 206–9
Storage shed, 157
Stringers, 151–52
Studs, 17–18, 21, 23, 96
   in gable ends, 113–14
   king, 87, 90–91
   lumber for, 81–82
   marking placement of, 86
   mounting boxes on, 168
   nailing Sheetrock to, 189
   nailing siding to, 120–21, 126
Styrofoam, 57, 70–72
Subfloor, 17, 23–24
   installing, 72–74
   plywood for, 58
   in sketch, 34
Surface wiring, 163

Tape measure, 7
Television cable hookup, 175–76
Template rafter, 107–8
Temporary slide stops, 131–32
Terminology, 7–9, 91, 149–50
Termite shields, 65–66
Tie beams, 20, 107
   installing, 112–13
Toenailing, 68–69
T-111, 116, 122, 125, 127, 139, 186, 204
   staining, 218

   weathered, 219
Tools, 5–8
   for spackling, 193
Top plates, 17, 96, 98, 100, 101
Treads, 149–52
Trim, 11
   exterior, 139–43, 219–20
   interior, 186, 198–213, 220, 221
Trimmers, 86, 87, 95
Trowels, 193
Truss plates, 53, 66

Underlayment, 58, 213
Unit run, 149–51

Vapor barrier, 72, 180–83
Varnish, 213, 222
Vinyl tile, 213, 216

Wallboard, see Sheetrock
Wall plates, 17
Walls, 11, 81–101
   cutting wood for, 85
   Eastern-style framing of, 98–100
   in frame design, 14, 15, 17–18
   insulation of, 181–83
   interior, 156, 181, 184–97, 220–21
   laying out, 83–84
   ordering lumber for, 81–83
   raising, 92–93
   sheathing and siding of, 115–17, 120–26
   and size of building, 29–30
   in sketch, 28
   squaring up, 100–101
   Western-style framing of, 88–92
Wall sills, 17
Western-style framing, 29, 87–92
Windows, 27, 75–80, 115

Windows (*cont'd*)
   framing, 86–87, 89, 90, 94–95
   installing, 117–20
   insulating around, 183
   lumber for openings, 82–83
   painting, 219
   in sketch, 28, 35
   trimming, 199–204
Wiring, 157, 160–76
   basic circuits, 168–76
   boxes, 167–68
   gauges and color codes for, 166–67
   surface, 163
Wood preservatives, 218–19
Wood stoves, 79, 157, 224–28
Woodwork, *see* Trim
Working sketch, 13

Zoning ordinances, 25